Google Glass™ FOR DUMMIES®
A Wiley Brand

by Eric Butow and Robert Stepisnik

FOR DUMMIES
A Wiley Brand

Google Glass™ For Dummies®

Published by: **John Wiley & Sons, Inc.,** 111 River Street, Hoboken, NJ 07030-5774, www.wiley.com

Copyright © 2014 by John Wiley & Sons, Inc., Hoboken, New Jersey

Published simultaneously in Canada

For general information on our other products and services, please contact our Customer Care Department within the U.S. at 877-762-2974, outside the U.S. at 317-572-3993, or fax 317-572-4002. For technical support, please visit www.wiley.com/techsupport.

Wiley publishes in a variety of print and electronic formats and by print-on-demand. Some material included with standard print versions of this book may not be included in e-books or in print-on-demand. If this book refers to media such as a CD or DVD that is not included in the version you purchased, you may download this material at http://booksupport.wiley.com. For more information about Wiley products, visit www.wiley.com.

Library of Congress Control Number: 2013954211

ISBN 978-1-118-82522-8 (pbk); ISBN 978-1-118-82519-8 (ebk); ISBN 978-1-118-82514-3 (ebk)

Manufactured in the United States of America

10 9 8 7 6 5 4 3 2 1

Contents at a Glance

Table of Contents

Introduction

Welcome to *Google Glass For Dummies,* a resource for using the Google Glass wearable computer in your everyday life.

Glass is one of the first devices in the wearable-computing market, and though that's really exciting — and probably is one of the reasons why you purchased Glass in the first place — you may not be prepared for the differences in the user interface. Interacting with Glass is unlike anything you've experienced before.

To help get you up to speed quickly, this book dishes out information about Glass in easily digestible chunks so that you can get answers to your questions fast and then get back to using your Glass to get directions, take pictures, and much more.

About This Book

Unlike other books that require you to read an entire chapter in one sitting to understand what's going on, this book is more like a dictionary — or, if you prefer, a printed version of what you find in online help. That is, you can search the book for the information you need, read the page that contains the answers, and then put the book down and get back to work (or play).

We discuss the following subjects (among others) in this book in plain English:

- ✔ Fitting Glass comfortably on your head and fitting the device with shades and shields

- ✔ Setting up Glass by using the MyGlass app in your web browser or on your smartphone

- ✔ Manipulating Glass by using your voice and your finger on the touchpad

- ✔ Using Glass responsibly so you can show it to others while respecting their privacy and safeguarding your data

- ✔ Taking pictures and shooting videos, and then sharing those photos and videos with others

- ✔ Navigating to your destination and getting suggestions from Glass about places to visit and photograph

- ✔ Troubleshooting and resolving problems with Glass

Within this book, you may note that some web addresses break across two lines of text. If you're reading this book in print and want to visit one of these web pages, simply key in the web address exactly as it's noted in the text, pretending that the line break doesn't exist. If you're reading this as an e-book, you've got it easy; just click or tap the web address to be taken directly to the web page.

Foolish Assumptions

We've taken the risk of making some assumptions about what you already have or know:

- ✔ You have some experience with computing technology, with a computer, tablet, and/or smartphone. You may have also seen Glass on the web or watched someone demonstrate the device.

- ✔ You either have Glass and want to know how to use it, or you have a keen interest in finding out more about it — perhaps to make a decision about purchasing the device.

- ✔ You want to know how to make Glass do what you want it to do.

We also assume that you'll want to have this book nearby to help answer questions or resolve problems in your time of need.

Icons Used in This Book

Just like the screen of Google Glass, several icons throughout this book indicate important information. These icons, which appear in the left margin, tell you something extra about the topic you're reading about or hammer a point home.

The Tip icon marks tips (duh!) and shortcuts that make using Glass easier.

The Remember icon marks information that's especially important to know. To siphon off the most important information in each chapter, just skim the Remember text.

The Technical Stuff icon denotes information of a highly technical nature that you can skip if you want to (but we hope you won't).

The Warning icon tells you to watch out! It flags important information that may save you headaches — not to mention your data — when you use Glass.

Beyond the Book

We've written a lot of extra content that you won't find in this book. Go online to find the following:

- ✔ **Web Extras:** www.dummies.com/extras/googleglass

  Here, you'll find online articles that go into more detail about how to use various Glass apps and discoveries we've made about Glass as we've used it in the real world.

- ✔ **Cheat Sheet:** www.dummies.com/cheatsheet/googleglass

  Our Cheat Sheet offers plenty of tips and tricks for making life with Glass easier.

- ✔ **Updates to this book (if any):** www.dummies.com/extras/googleglass

  Google continually updates the Glass hardware and software, so you can keep your book up to date by reading our update articles.

Where to Go from Here

This book is yours, so you can annotate and augment the text any way you want — by highlighting text, adding notes, or placing bookmarks at strategic locations so you can return to those places quickly.

If you've just purchased Glass and want to get grounded in what the device is all about, flip to Chapter 1. If you can't wait to get your Glass out of the box and start playing with it, however, start with Chapter 2.

Part I
Getting Started with Google Glass

In this part . . .

- ✔ Find out what Google Glass is and what you can do with it.
- ✔ Get familiar with the Glass device, including how to fit it on your head properly.
- ✔ See how to charge Glass when the battery runs low.
- ✔ Set up Glass with a browser and on your Android smartphone by using the MyGlass app.
- ✔ Visit www.dummies.com for great Dummies content online.

Chapter 1

The World Through Glass

Once a new technology rolls over you, if you're not part of the steamroller, you're part of the road.

— Stewart Brand

Do you remember the first time you used your smartphone? Your tablet? What about your first computer? Your experience each time may have been the same: Not only are you using something that's exciting and new, but also, your life is changing for the better.

That's probably the same feeling you have with Google Glass, which is the next step in mobile computing. Instead of having to fumble with your smartphone to get things done, now you can wear a computer on your head and control it (mostly) with your voice.

Google Glass isn't an immersive experience, in which you keep the unit on all the time and you're continually distracted by it. Glass is on only when you want it on. Indeed, the Glass screen is off by default, and just as with any computer or smartphone, you have to take specific actions to turn it on.

Google Glass is a form of *augmented reality,* which means that it provides a live view of your physical environment augmented by what you see on the screen, such as a map of your current location. Augmented-reality environments cover your entire field of vision. Google designed the Glass screen to be above your field of vision, however, to make it as unobtrusive as possible.

What Is This Thing Called Glass?

Wearable devices are a hot new area of computing, and companies are scrambling to create computers that you can wear on your wrist (called *smartwatches*) and computers that sit on your head like eyeglasses (called *wearable computing glasses*).

In the case of wearable computing glasses, Google recognized that seeing is the most natural way of consuming information and that computing technology has become small enough for a computer to fit within the frame of a pair of glasses.

Google created a small screen that pairs information with what you see in the natural environment, as shown in Figure 1-1. What's more, it developed an operating system that understands your voice commands so that you can perform tasks without using your hands, arms, or neck.

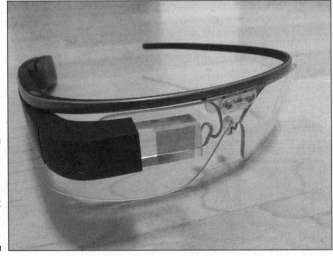

Figure 1-1: The screen and camera component of Google Glass.

Thus, Google Glass and the market niche of wearable computing glasses were born. Glass was a media sensation from the time it was announced, but despite all the hype (both good and bad), the mission of Glass hasn't changed: to deliver information as quickly and conveniently as possible with cutting-edge technology.

Google Glass comes in five frame colors — charcoal, tangerine, shale, cotton, and sky — so you can match your Glass to your personality.

Google provides clear and shaded protective shields with Glass. The clear shields protect your eyes from the wind and other elements, and also make your Glass look like any other pair of glasses. The shaded shields reduce ambient light so you won't squint outside on a sunny day.

At this writing, prescription shields have also been announced for Glass. You can search for *Google Glass prescription* to see what prescription lenses are available.

Before you buy, talk with your optometrist to find out whether using Glass with prescription lenses is right for you.

What does Google Glass do?

Here's what Google Glass does when it's on and connected to the Internet:

- Takes photos and videos, and sends them to one or more of your contacts. The Glass camera sees the world through your eyes at the very moment you take the photo or record the video.

- Sends e-mail and text messages to your contacts, and receives the same from them.

- Allows you to chat live via video with one or more Google+ friends via Google Hangouts.

- Sends and receives phone calls.

- Searches the web with the Google search engine (of course) so that you can find information easily.

- Translates text from one language to another. Glass speaks the translated text and also shows a phonetic spelling of the translated word(s) on its screen.

- Provides turn-by-turn navigation with maps as you drive, ride, or walk to your destination.

- Shows current information that's important to you, including the time, the weather, and your appointments for the day.

- Recognizes the song that's playing on the device and identifies the artist(s) singing the song, in case you don't know.

If your Glass is on but not connected to the Internet, you can only take photos and videos with the device. If you have a Wi-Fi connection, functionality is expanded to sending and receiving e-mail messages, connecting to the Internet, sharing information on social networking websites, and participating in Google Hangouts for videoconferencing. You need to connect Glass to your phone if you want to use your phone's 3G or 4G Internet connection, send SMS messages, and use GPS (Global Positioning System) so Glass can find your current location.

How does Glass differ from other mobile devices?

Google Glass competes not only with other wearable computing glasses, but also with other mobile devices, including smartphones and smartwatches. Yet Glass is the most convenient wearable device produced to date and gives you the broadest access.

With Glass, for example, you don't need to move multiple parts of your body to fish for your smartphone, keep the phone in your hand, and then hold your arm up to your ear for a while. Those maneuvers can be more painful than just moving your head and using your voice with Glass. What's more, you can have a live video conversation with someone else, and that other person sees what your Glass camera sees.

In the case of a smartwatch, you have to move your wrist to your mouth (and possibly roll up your sleeve) so you can talk into the phone or see what's happening on the screen. Smartphone screens are relatively small, and smartwatch screens have to be even smaller to fit on your wrist, so your eyes may be strained. The Glass screen looks like a 25-inch television screen that's 8 feet away from you, so you don't strain to see things.

You'll still be talking on Google Glass in public as you would on a smartphone or smartwatch, of course, and you should follow proper etiquette when using your Glass. You find out about following Glass etiquette and using the device responsibly in Chapter 7.

How Glass Fits into Your Life

Several Glass applications (*apps,* for short) have been developed to make users' lives easier:

  ✔ **SMARTSign:** This app teaches American Sign Language (ASL) to Glass wearers. The app was designed to help parents communicate with their hearing-impaired or deaf children and also teach ASL to their other children who can hear. Visit the SMARTSign website at www.cc.gatech. edu/content/smartsign-google-glass.

- ✔ **Field Trip:** This app for Glass displays information about a nearby point of interest. If you're near a landmark, you see information on your Glass about that landmark. You can also find out about destinations you're thinking about visiting, such as a new restaurant in town. Yes, you can do the same things with your smartphone, but the process is more cumbersome than seeing the information about your landmark or destination as you're looking at it in real time. You can get more information and activate the app at www.fieldtripper.com/glass.

- ✔ **Question-Answer and Memento:** Through its OpenGlass project, Dapper Vision has developed two apps to help the visually impaired and blind "see" what Google Glass is seeing. The Question-Answer app allows the user to take a picture of what his or her Glass is looking at and then sends the picture to a social networking site such as Twitter. When someone responds with a description of the picture, Glass reads that response to the user. The other app, Memento, allows a user to record voice commentary about a specific scene and then plays the commentary when a visually impaired user wearing Glass visits the scene. You can get more information about these apps at www.openshades.com.

More Glass apps are being developed daily. You find out more about available Glass apps and shopping for apps in Chapter 12.

The catalog of apps is always changing. By the time you read this book, some of the apps we discuss may have different names; may have added or changed functionality; or may not be offered anymore, either because they weren't successful or were replaced by newer versions. Be sure to visit the apps' websites and the MyGlass website (www.google.com/myglass). You find out more about MyGlass in Chapter 5.

What You Can Do with Glass

This section provides some examples of what you can do with Glass. Some of them are hypothetical, based on what Glass can do now; others are based on real-world experiences.

Manage air travel

A married couple walk through an airport. They don't have to look at the television screens displaying flight information because they're both wearing Glass, so they both get real-time flight updates on their Glass screens. When they eat at an ethnic restaurant in the airport, they look up ways to translate phrases such as *good* and *thank you* into the proprietor's native language.

When the couple finish eating, they see a map of the airport on their Glass screens, showing them the correct boarding gate. During the flight home, the wife speaks a quick message to her mother about their time of arrival, and that message appears on her mother's phone. The couple didn't have to add anything to their devices to make all this work; all that functionality comes with each Glass out of the box.

Do your job more efficiently

You can use Google Glass to collaborate with other professionals and improve your job results, as follows:

- A surgeon at Ohio State University used the video features of Google Glass to show live knee surgery to other doctors connected through a Google Hangout. This approach provides an opportunity not only to teach, but also to collaborate live with other surgeons while a procedure is performed. In both cases, viewing doctors can see what's happening just as the performing doctor sees it.

  Doctors can also use Glass to get updated information such as the latest CT (computed tomography) scan results, reports from specialists, and drug interactions.

- Mutualink (www.mutualink.net) has developed an app that allows public-safety personnel and first responders to use Glass to communicate with colleagues, get information about patients, and keep abreast of emergency situations. Emergency medical technicians who use this app on Glass can get a list of allergies onscreen while keeping their hands free, as well as communicate with an emergency-room doctor before the patient arrives at the hospital.

- Fiberlink (www.maas360.com) has created mobility management software that helps information technology professionals find missing smartphones, police the company network, and change network user information from Glass. If you're often on the go but need to keep tabs on a network, using Glass with Fiberlink software may be a good solution.

Get fit and get cooking

The LynxFit app (www.lynxfit.com) contains a wide variety of workouts (including yoga, CrossFit, running, and walking) so you can follow the workout regimen on your Google Glass screen as you exercise. What's more, the app tracks your statistics as you exercise so you can see just how many calories you're burning — and perhaps get motivated to work harder.

If you like to cook but want to use both hands for cooking and/or don't want to get your smartphone or tablet smudged with food stains, you can use the Allthecooks app (www.allthecooks.com). This app lists recipes with step-by-step instructions so you can cook delicious meals successfully. The instructions appear on the Glass screen, so you don't have to wash your hands to consult a smartphone or tablet; as you complete each step, simply tell the Allthecooks app to proceed to the next one.

Meet people

You have a built-in conversation starter when you wear your Glass out in public. Glass gets attention wherever it goes, whether it's at the mall, on the street, or in the store. Also, don't be surprised if strangers want to try on your Glass. For more about how to handle questions and requests, see Chapter 7.

Is Glass the Most Exciting Thing Ever?

Glass is exciting because it's the next step in the evolution of both computing and human communication. The amount of excitement you feel may vary, depending on what you use Glass for and perhaps depending on the reactions you get from others. But Glass is only going to get better, because the device receives regular operating-system updates, and companies are continuously developing new apps for it.

With all this in mind, it's time for you to begin your adventure with Glass by taking the device out of its box and setting it up as we show you in the next two chapters.

The downside of Glass

Like any piece of new technology, Glass has a few limitations. Here's what we know at this writing:

- **Work in progress:** Google calls Glass a ten-year commitment, stating that it'll be ten years before the device gets to where Google wants it to be. The company is depending on users to provide feedback that will make Glass better, so there will be some fits and starts as Google tries to determine how best to balance Glass features with users' concerns. While this book was being written, for example, Google removed facial-recognition software from Glass because of privacy concerns.

- **For teenagers and adults only:** Google recommends that children under 13 not use Glass, as it could harm developing vision.

- **A brief return period:** You're entitled to return your Glass if you decide that it isn't for you. You should do that during the applicable refund period, which is 30 days at this writing. Google may also charge you a restocking fee for your return.

- **A caveat if you had Lasik eye repair:** For those of you who had Lasik surgery to repair your vision, get advice from your optometrist about using Glass.

- **Future ads:** Google has patented a "pay-per-gaze" advertising model that could bring advertising to Glass when the number of users reaches critical mass. It remains to be seen what form that advertising will take, but ads are likely to come to Glass sooner rather than later.

- **No support for forms:** Glass doesn't support the use of forms to sign in to websites. You can't even use a different device to log in and then use the website on Glass. Therefore, if you plan to use a website that requires a login form, be sure to use that website on another device, such as a computer, tablet, or smartphone.

- **No support for Adobe Flash:** Flash videos won't play in the Browser app (see Chapter 9), so you won't be able to view a lot of videos on YouTube and other popular video-sharing websites. YouTube and similar websites are now using videos produced with HTML5, however, and you can view those videos on Glass without any trouble.

Some (or even all) of these problems may be resolved by the time you read this book. YouTube is owned by Google, for example, so it's reasonable to assume that Google wants Glass users to be able to view videos on the device. Bookmark the Google Glass website (www.google.com/glass) on your Glass, and check it regularly to get the latest news and updates.

Chapter 2

Finally On Your Head

In This Chapter
▶ Getting to know the device
▶ Powering on your Glass
▶ Charging the battery
▶ Fitting yourself with Glass

In order to change the world, you have to get your head together first.

— Jimi Hendrix

*W*hen you finally have Google Glass in your hands, you'll want to put it on your head right away. It's an amazing feeling when you first look up and see a small screen that lights up and waits for your orders. You have two choices about what to do next: Look at the clock on the first screen, or start exploring what this device is capable of doing.

Although watching numbers change on the clock can be exciting, there's something even more exciting to explore just above your line of sight: a computer that performs almost every function you can imagine.

This chapter discusses how you can use Glass comfortably in your everyday activities and look good while wearing it.

Put Your Glass On

Before putting your Glass on your head, you should familiarize yourself with all the features that the device offers. We know that some of you are smart enough to figure things out on your own, but if you're not a technical type, this section is for you. Here, we walk you through unboxing Glass, reviewing all the Glass features, and turning your device on for the first time.

Unboxing Glass

To use Glass, you have to take it out of its beautiful packaging so that you can get acquainted with everything that's inside. You should find the following items inside the large box:

- ✔ **Glass:** Don't be surprised if it's not there anymore, because if you're excited as we were when we first opened the box, you may already have the device on your head.

- ✔ **Micro USB cable:** You use this cable to connect your Glass to a computer or a wall charger.

- ✔ **Wall charger:** The charger is paired with the micro USB cable. We talk about charging later in this chapter.

- ✔ **Pouch:** The pouch serves to protect your Glass and its accessories. It's a really nice gesture by Google to include a light, durable case that makes Glass safe.

- ✔ **Earbud:** An *earbud* is a miniature headphone that fits inside the ear. The earbud for Glass is attached to a piece of cardboard and placed inside the pouch. Place the earbud in your desired ear and then connect it to Glass by inserting its plug into the micro USB port, which is the same port you use to charge Glass.

- ✔ **Additional nose pads:** Everyone is unique, and these nose pads adapt to uniqueness. Try them all to see which pair fits you best.

- ✔ **Q&A:** Read this list of common questions and answers about Glass if you're curious.

Your package should also include two smaller boxes. Google wants you to take your Glass outside, so the device comes includes a pair of eye-protecting lenses, called *shields:*

- ✔ **Glass Shade:** This tinted shield is Google's replacement for your regular sunglasses, designed to deliver the best experience on sunny days. Now go play outside.

- ✔ **Glass Shield:** This clear shield offers great protection on windy days.

A case is included with the shields so you can store them while they're not in use. You can change shields easily. Just put the shield you want to use between the nose-pad holders, and you're set. For more about changing shields, see Chapter 11.

Getting to know your Glass

To start mastering something, you must understand the basics. Figure 2-1 shows the basic key features of Glass:

- ✔ **Display:** The display shows you all the information that's available on your Glass, including e-mails, pictures, videos, and any other content.

- ✔ **Camera button:** Press and release this button to take a photo. If you hold the button down, Glass starts to record a video. You can find more about taking photos and videos in Chapter 8.

- ✔ **Touchpad:** On the right side of Glass is a flat surface called the *touchpad,* which allows you to navigate among the screens you see in the display. Swipe forward and backward to navigate the Glass timeline, which displays your current and past activities, or swipe down to return to the previous screen.

  You can also start Google voice search, which allows you to search the web. Tap the touchpad three times slowly when the display is off, and speak your search terms into your Glass. If you're at the Home screen, you can start voice search by saying "OK Glass, Google . . ." followed by your search term. Chapter 9 talks more about searching with Glass.

- ✔ **USB/power connector:** The USB/power connector is on the bottom of the Glass unit, near the end of the touchpad. Attach the USB cable to charge Glass, or attach the power connector to connect your Glass with your computer to exchange files.

- ✔ **Power button:** On the inner right side, near the end of the touchpad, is a button that turns your Glass on and off. Press the button to power your device on or shut it down.

- ✔ **LED:** The LED, located on the back of the touchpad, displays white light when Glass is charging, starting up, or shutting down. The LED also blinks slowly while you're charging the device.

- ✔ **Speaker:** This oval speaker, engraved with the Glass logo, delivers all the sounds from the device.

- ✔ **Battery:** At the bottom-left side of the frame is the battery that powers your device. See the "Charging Glass" section later in this chapter to get more information about the health of the battery.

- ✔ **Frame:** The Glass frame is extremely durable and can be bent to fit any face. Try it.

- ✔ **Nose pads:** Those little things in the middle of the frame are there to keep your Glass from falling off your head. Feel free to adjust the pads to suit your needs. For more information, see "Fitting Your Glass" later in this chapter.

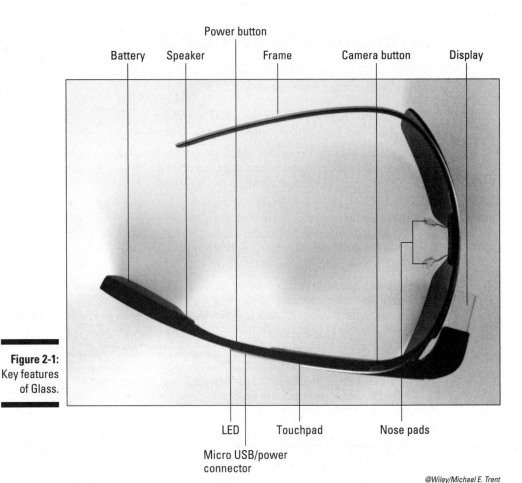

Battery Speaker Power button Frame Camera button Display

LED Touchpad Nose pads

Micro USB/power connector

@Wiley/Michael E. Trent

Figure 2-1:
Key features
of Glass.

Turning Glass on and off

The first time you turn on a new device, you get a warm feeling that some-thing new is happening and that you're a part of it. So do this:

1. **Press the power button firmly.**

    The LED behind the button turns on and stays on while Glass starts up. After the device boots successfully, the display turns on (see Figure 2-2), and you hear a rising chime sound.

Figure 2-2:
The first
screen you
see after
powering
on.

9:15

" ok glass "

2. **To turn off your Glass, press the power button until you hear a falling chime sound.**

   That's it.

Whenever Glass is turned on, you can set either of these options at any time:

✔ **Head Wake Up:** This option is on by default. When you tilt your head up at a specified angle, your Glass turns the screen on and waits for your orders. For more on this feature, check out Chapter 6.

✔ **On-Head Detection:** This feature tells Glass whether or not you're wearing it. Glass turns off when you take it off your head so that nothing bad will happen if you accidentally tap the device while you're giving it a break. For details on configuring On-Head Detection, see Chapter 6.

If your Glass isn't responding to your commands, you can fully power it down (that is, perform a hard reset) and then restart it by holding the power button down for 15 seconds. After Glass shuts down, restart the device by pressing the power button again.

You may also see a low-battery warning — perhaps the most annoying message of all on any battery-powered device. Until humans invent a power source that provides an infinite supply of energy, you'll see a message similar to Figure 2-3 every time Glass gets hungry. When Glass runs out of battery power, it powers itself down automatically.

Fear not — this shutdown is normal behavior. Just connect your charger to Glass (see the next section). When the device has enough power, it turns on again automatically.

Low battery. Connect charger.

Figure 2-3:
Low-battery
warning.

Charging Glass

Every device on this planet has to have energy to do its thing, and Glass is no exception. Fortunately, the kind people at Google made the charging process really easy for you. The box that Glass comes in (refer to "Unboxing Glass" earlier in this chapter) contains a funny-looking micro USB cable and a charger. The charger and the cable plug are two-tone — black and white — so that you always know which way to plug the USB connector into the charger.

Glass comes from the factory partially charged so that you can start using it right away. Before using your Glass for the first time, however, you should charge it fully. (You don't want to see a low-battery message while you're in the middle of exploring your device.) Just follow these three simple steps:

1. **Connect the micro USB cable to the wall charger.**

   Take the narrow end of the cable and plug it into the wall charger.

2. **Connect the free end of the micro USB cable to your Glass.**

   Flip your Glass over and find the socket that matches the free end of the micro USB cable, above the power button. The white part of the micro USB cable should face the outside of the Glass device.

3. **Plug the wall charger into a wall socket.**

   When you've connected everything successfully, the LED near the power button starts blinking slowly, and the display shows an animated battery icon. You're officially charging your Glass for the first time.

Another way to charge Glass is to connect it to your computer. The downside of this process is that it takes longer than the cable-and-wall-charger method.

You can use your Glass while the battery is getting its life back. The micro USB cable is designed so that it won't be in your way while it's connected to Glass.

If you're charging while you're within range of a Wi-Fi network that you're already using for your Glass and any other Wi-Fi device (such as your smartphone), Glass automatically checks for software updates. It also synchronizes all the pictures and videos you've stored on it with your Google+ account so that you can share those photos and videos with your Google+ friends.

The Glass battery isn't removable, so if the battery dies, the device isn't useful anymore. The best thing to do in this case is contact your Glass Guide, who can help you get a replacement device. You can reach your Glass Guide by phone at (800) 452-7793 or by e-mail at glass-support@google.com. You find out more about Glass Guides in Chapter 17.

If you find that your battery isn't staying charged very long, Glass Guides suggest that you plug the device in for an hour and then unplug the device for an hour. You'll perform this plug-and-unplug regimen 12 times within a 24-hour period. When you're finished, use Glass as you normally would, and see whether it stays charged as long as you expect. If not, contact a Glass Guide (see the preceding paragraph).

Fitting Your Glass

You have two good reasons to read this section:

- ✔ You should feel comfortable while wearing your Glass.
- ✔ You'll probably attract a lot of attention, so it's important to look good.

The fitting process is simple, and if you fit the device on your head correctly, fitting should be a one-time event. The key to a comfortable Glass experience is making sure that the device isn't in your way when you're enjoying your everyday activities.

Here are a few pointers for wearing Glass comfortably and looking great while doing it:

- ✔ **Position the display above your line of sight.** The small cube that functions as the display (refer to Figure 2-1 earlier in this chapter) should be located above your eye, not directly in front of it. Glass is designed to stay out of your way when it's not needed.

✔ **Adjust the display as needed.** Move the display forward or backward until you're able to see every corner of the screen clearly.

✔ **Adjust the nose pads.** Like every other pair of glasses you may have worn, Glass has to fit perfectly on your nose to work properly. The nose pads are adjustable, so do a little tweaking to get the best fit for your head.

It takes time to get used to wearing Glass, just as it does to get used to wearing a pair of new eyeglasses. Have fun exploring!

Chapter 3

Setting Up Your Google Glass

● ●

● ●

I like control.

— Michael Jordan

Imagine a life without settings. You can pick up any device that you want to use, and because it knows who you are, it adapts to you instantly.

Now return to the present day. You have your shiny new device in your hand (or, in this case, on your head), and you want to start using it immediately. First, though, you must make a bunch of settings to make it usable.

That process could be tricky, but fear not; this chapter shows you how everything works.

Completing the Google Glass Checklist

To keep your Glass connected to the Internet and/or to your Android smartphone, you need the following:

✔ **Glass:** Make sure that you have it charged up and ready to go (see Chapter 2). You don't want to be surprised by news of an empty battery.

✔ **Google account:** A Google account is required to set up Glass. If you don't already have a Google account, you can create one at `http://accounts.google.com`.

Your Google account should have Google+ and Gmail services enabled. These two services are required so that Glass can authenticate you and use Google+ features.

✔ **Connection to the Internet:** Internet connection is essential for Glass. Without it, Glass is just an expensive pair of glasses with a great camera.

✔ **Computer with web browser:** Your primary computer will do just fine, as long as it has a working Internet connection and a web browser.

✔ **Smartphone or tablet with Bluetooth:** A Bluetooth-enabled device allows you to use some of the phone features of your Glass, such as making phone calls, sending and receiving text messages, and getting directions to your next destination. You can also use your smartphone carrier's data connection to connect your Glass to the Internet.

✔ **MyGlass app:** If you have an Android smartphone, you should install the MyGlass app to configure Google Glass.

Because Google produces both Glass and the Android operating system, it should come as no surprise that if you use Glass with an Android smartphone, you can take advantage of some features you can't get when you use Glass with an iPhone. For that reason, we cover the Android process in this chapter. You find out more about downloading and using the MyGlass app in "Setup on an Android phone" later in this chapter.

Setting Up Your Glass in Many (OK, Two) Ways

When you put Glass on your head and power it on for the first time, you see the Glass logo screen, followed by a welcome screen. The welcome screen asks you to start the setup process by swiping forward on the touchpad.

After you swipe forward, you see a screen that says `Great`, which means that Glass acknowledges that you know how to swipe forward on the touchpad. Start the swiping tutorial process by swiping forward on the touchpad once more. Then follow these steps:

1. **Respond to the tutorial's animation and onscreen request (`Let's practice swiping down`) by swiping down on the touchpad.**

   A new screen says `Perfect` and asks you to swipe forward.

2. **Proceed to the next step by swiping forward on the touchpad.**

   The next screen features an animation that shows you how to tap the touchpad.

3. **Tap the touchpad to go to the next step.**

   The next screen shows a menu that asks how you want to set up your Glass (see Figure 3-1).

Figure 3-1:
The three
setup
options.

4. **Tap the setup option you want.**

5. **Do one of the following, depending on the option you chose in Step 4:**

   - To use an Android smartphone, skip to "Setup on an Android phone" later in this chapter.

   - To use your computer, proceed to "Setup with a web browser" later in this chapter.

Setup on an Android phone

If you have an Android smartphone, you can set up your Glass with the MyGlass app. MyGlass also lets you use additional features, including GPS, SMS messaging, and screencasts. You can find more about these features in Part III of this book.

To install MyGlass on your Android phone, your phone needs to be running version 4.0.3 (Ice Cream Sandwich) or later of the Android operating system.

Installing MyGlass

To install the MyGlass app on your Android smartphone, follow these steps:

1. **Open Google Play from the main Home screen or the All Apps screen on your phone.**

2. **Tap the magnifying-glass icon at the right end of the menu bar at the top of the screen.**

   The Search box appears in the menu bar.

3. **Type MyGlass in the search box and then tap the Search button.**

   The MyGlass app should be the first item listed in the resulting screen.

4. **Open a description screen for the app by tapping MyGlass.**

5. **Tap the Install button in the top-right corner of the description screen.**

   You see the install screen, which describes the requirements for using the MyGlass app.

6. **Tap the Accept & Download button.**

   The MyGlass app downloads to your phone.

Configuring MyGlass for your Glass

When the MyGlass app is installed on your phone, you see the green Open button on the Google Play screen. Follow these steps to run and configure it:

1. **Tap the Open button to run the MyGlass app.**

2. **In the screen shown in Figure 3-2, tap the Play icon to view the introductory video.**

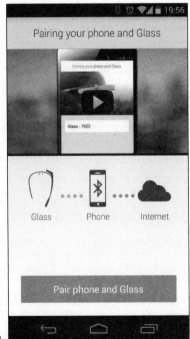

Figure 3-2: Watch the introduction video to see how to fit Glass on your head properly.

3. **When you finish viewing the video, or to skip viewing the video, tap Continue.**

4. **Do one of the following:**

   • If you have more than one Google account, select the Google account that you want to use with your Glass.

   • If you have only one Google account, proceed to Step 5.

5. **Tap Pair Phone and Glass (refer to Figure 3-2).**

   If the Bluetooth connection is off on your smartphone, MyGlass turns on Bluetooth automatically.

6. **In the resulting screen, tap the Glass device (see Figure 3-3).**

   You see a Bluetooth pairing request screen on both your smartphone and your Glass.

Figure 3-3:
Tap your
Glass to pair
it with your
smartphone.

7. **On your Glass, tap the touchpad to pair Glass with your smartphone.**

8. **On your smartphone, tap Pair in the Bluetooth Pairing Request pane (see Figure 3-4).**

   After a short while, the MyGlass app finishes pairing the smartphone and the Glass device. Then your Glass plays a chime through the speaker and displays a welcome message that includes your Google profile picture.

9. **In the next screen on your smartphone, get information about the Glass connection by tapping the Device Info card (see Figure 3-5).**

10. **Begin exploring Glass and the MyGlass application on your smartphone.**

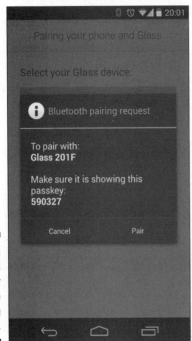

Figure 3-4:
Pair Glass
with your
smartphone
by tapping
Pair.

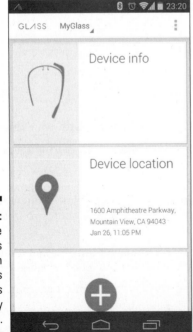

Figure 3-5:
The
MyGlass
app screen
shows
where Glass
is currently
located.

Setup with a web browser

You can set up Glass in a web browser on your computer, and you'll get the best results from the most recent version of your browser of choice.

If you're uncertain which web browser to use, we recommend Google's Chrome. Chrome goes great with Glass because it's also a Google product. You can get it at www.google.com/chrome.

Here's how to set up your Glass in your web browser:

1. **Visit www.google.com/myglass in your computer's web browser.**

2. **Sign in, using your Google account ID and password.**

   You see the screen shown in Figure 3-6.

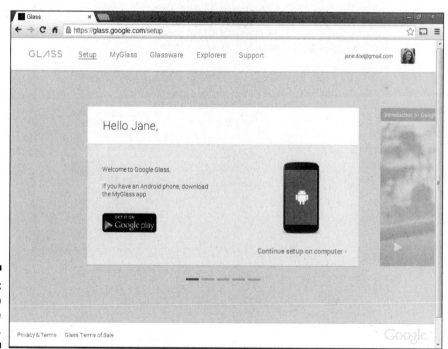

Figure 3-6:
The setup
welcome
screen.

3. **Click Continue Setup on Computer.**

   The Introduction to Glass page appears in the browser window, and an introductory video plays onscreen, as shown in Figure 3-7.

   If you want to view the introductory video later, you can do so by visiting https://www.youtube.com/watch?v=cAediAS9ADM.

4. **Move to the next screen by clicking Continue.**

5. **On the next page (see Figure 3-8), click the check box titled I Agree to Glass Terms and Conditions; then click Continue.**

   The Setup Wifi page appears in the browser window, and an instructional video plays (see Figure 3-9).

   If you want to view the instructional video later, visit https://www.youtube.com/watch?v=g3ncmeGaKN0.

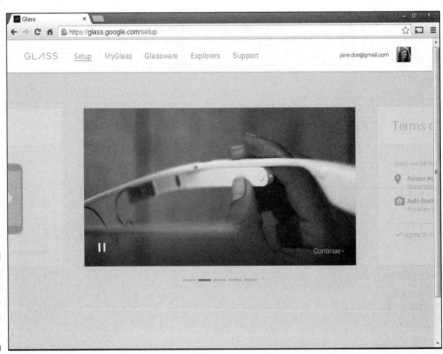

Figure 3-7: The Introduction to Glass video.

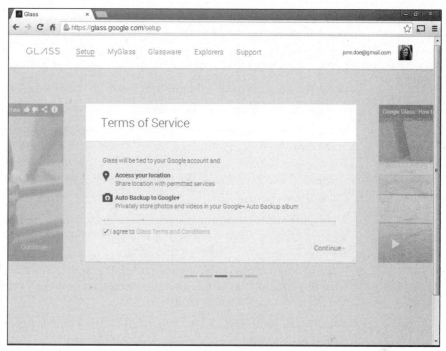

Figure 3-8:
The Terms
of Service
page.

Figure 3-9:
The Setup
Wifi page
for a
computer.

6. **Move to the next screen by clicking Continue.**

   The Connecting to Wifi screen appears (see Figure 3-10). This screen lets you sort out how you want to set up your Wi-Fi network so that your Glass can connect to the Internet.

7. **Enter the name of your Wi-Fi network and (if required) a password.**

   The network password is case-sensitive.

   To connect via a hidden Wi-Fi network, click the words *encryption type* in the text `Have a hidden password? Specify your encryption type` below the Network Password field. Enter your hidden Wi-Fi network information in the resulting screen and then continue to Step 8.

8. **Click Show QR Code.**

   A QR code appears in your browser (see Figure 3-11), and a small box appears on your Glass screen. (For more information on this code, see the nearby sidebar "What is a QR code?")

9. **Looking at your computer display, align the box on the Glass screen with the QR code on your computer screen so that your Glass captures the QR code, as shown in Figure 3-12.**

   You should see the text `Signing In`. Shortly after that, you see a welcome message that includes your Google account photo.

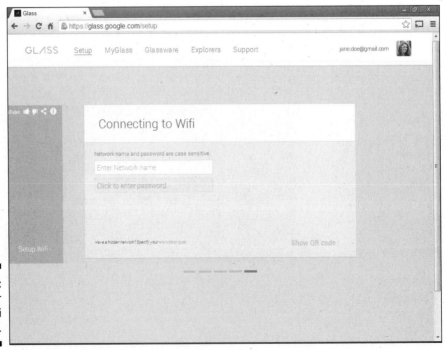

Figure 3-10:
Add your
Wi-Fi
network.

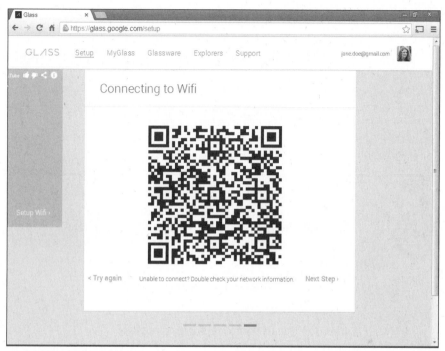

Figure 3-11:
The QR code for Glass to read.

Figure 3-12:
Align the center of the box on the Glass screen with the QR code.

What is a QR code?

A *QR code (QR* stands for *Quick Response)* is a two-dimensional bar code that phones and other devices with the appropriate reader can decipher. The code can store various types of data, such as URLs, links, geographic coordinates, and text.

You can download a QR reader from Google Play; find one by searching for *QR readers.*

If you can't log into your Wi-Fi network on your Glass, click Back in your browser and then click Try Again in the Connecting to Wifi page (refer to Figure 3-11). Verify your Wi-Fi connection and/or correct any mistakes.

10. **Click Continue.**

    The final setup screen appears (see Figure 3-13).

11. **Close the browser.**

    You don't need your computer for now.

12. **Tap the Glass touchpad and start exploring your Glass on the MyGlass website (see Chapter 5).**

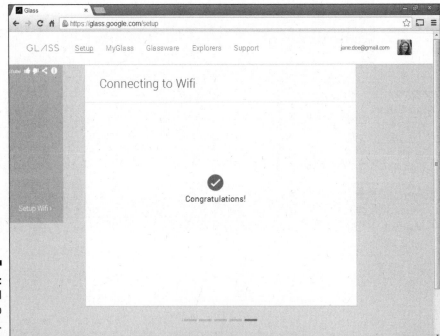

Figure 3-13:
The final setup screen.

Part II
Exploring Glass

See how to capture a Glass screen and display it on your desktop computer at www. dummies.com/extras/googleglass.

In this part . . .

✔ Discover how to use your voice and your hands to master your Glass.

✔ Find out how to view and manipulate cards in the timeline.

✔ Connect to the Internet and manage Glass apps via the MyGlass website and Glass settings.

✔ Figure out how to use Glass responsibly and share it with others safely.

✔ See how to capture a Glass screen and display it on your desktop computer at www.dummies.com/extras/googleglass.

Chapter 4

Mastering the Basics

Any fool can know. The point is to understand.

— Albert Einstein

*T*he fun begins when you get Glass to do what you want, and we show you how in this chapter, starting with the basic gestures you need to make Glass go.

You also see how to navigate the timeline, which contains notifications about various activities. Your Glass notifies you about new activities as they occur, such as when you receive a new e-mail message.

Starting with the Home Screen

The Home screen, shown in Figure 4-1, is the first thing you see when you activate your Glass. To wake up Glass, tap the touchpad or lift your head up as though you're looking at the moon. Your Glass display turns on, and you see a screen displaying the current time and the text "ok glass".

If you're unsure where everything is on your Glass, see Chapter 2.

Figure 4-1:
The Home
screen.

Using Your Hands to Control Your Glass

You must perform certain basic operations every time you wear your device. The learning curve isn't flat, but it's not as steep as you may think. All the basic gestures are done on the flat surface on the right side of the device, called the *touchpad*. Some of the gestures are similar to those used on all other touchscreen devices.

The following sections make you a master of swiping, tapping, and other cool gestures.

Swiping

When you swipe on your Glass, you're moving objects on its screen. To swipe, just put your finger on the touchpad and drag it to another spot (see Figure 4-2).

You can swipe forward, backward, up, and down. As you swipe, objects on the screen move in the direction in which your finger is moving. If you want to scroll quickly through a long list of notifications, for example, swipe quickly on the touchpad. (We cover notifications, called *cards*, later in this chapter.)

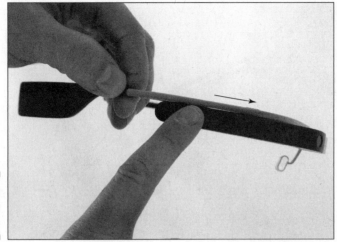

Figure 4-2:
The swiping
gesture.

@Wiley/Michael E. Trent

Tapping

Tapping means placing your finger on the touchpad to confirm an action onscreen. Think of it as the equivalent of pressing the Enter key on your computer keyboard or, if you prefer, tapping the Enter key on your smartphone or tablet.

If you use a smartphone, selecting or confirming items on the Glass screen comes naturally. All you have to do is tap the touchpad.

Dismissing an action

You can cancel an action by swiping your finger from the top of the touchpad to the bottom, as shown in Figure 4-3. Google calls this action the *dismiss* gesture. If you use the dismiss gesture while you're on the Home screen, for example, Glass turns off the display.

Apply this gesture if you want to move back one step anywhere on the timeline (see "Touring the Timeline" later in this chapter).

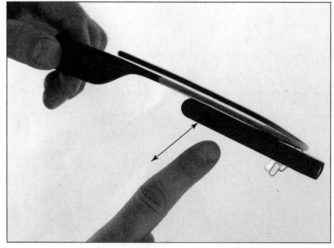

Figure 4-3:
The dismiss
gesture.

Waking up Glass

If you don't see anything on the Glass screen after you put the device on your head, and you're sure that you've turned it on, your Glass is in sleep mode. There are two ways to wake Glass up and show information on its screen:

✔ Tap the touchpad.

✔ Tilt your head up.

If you don't want to tilt your head up to wake up your Glass, you can change the wake-up setting so that you can wake Glass only by tapping the touchpad. For more information about changing settings, see Chapter 6.

Using Your Voice to Dictate to Your Glass

Though some teenagers may beg to differ, perhaps the most natural way to communicate with someone is to speak to him or her. Google designed Glass so you don't have to use the touchpad if you don't want to; you can just speak commands to your Glass.

When you start Glass and view the Home screen, the "ok glass" text is there to encourage you to speak "OK Glass" aloud. After you say "OK Glass," the Glass microphone recognizes your command and opens a menu of voice-command options (see Figure 4-4).

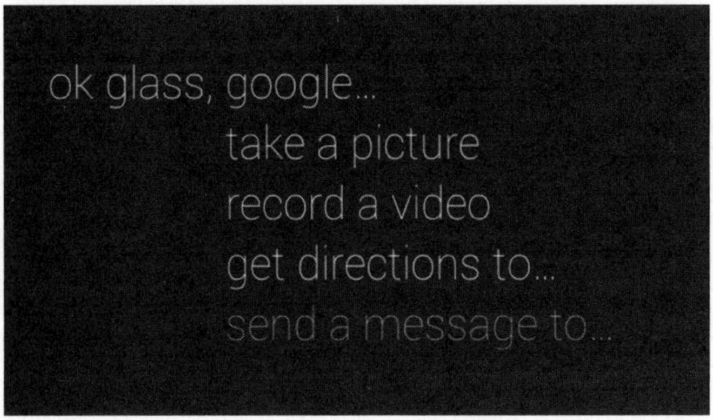

Figure 4-4:
The voice-
commands
menu.

You can move through the menu by tilting your head up and down. As you move your head, Glass moves the menu of voice commands to follow your head movement. When you're ready, say "OK Glass" again and speak a command from the menu, and Glass responds.

Voice commands include the following:

- ✔ **"Google":** Glass performs a Google search. You could say, for example, "OK Glass, Google current time in San Francisco."

- ✔ **"Take a picture":** Glass immediately takes a picture of what appears on its screen, adds the photo to a card, and places that card on the time-line. This voice command may be the one you use most often. For more about taking pictures, see Chapter 8.

- ✔ **"Record a video":** Glass records a ten-second video of what appears in front of the camera. You can extend that time by tapping the touch-pad and then choosing Extend Video from the menu or by pressing the Camera button.

- ✔ **"Get directions to. . .":** If you're using your Glass with a smartphone that's running the MyGlass app (see Chapter 5), you see turn-by-turn directions to your destination from your current location. Saying "OK Glass, get directions to San Francisco," for example, returns a screen with directions to The City. Find out more about navigating to a desired location in Chapter 11.

- ✔ **"Send a message to. . .":** If you're using an Android smartphone with your Glass, this command sends an instant message to the phone number that's associated with the name of the contact. (If you're using an iPhone or a Wi-Fi connection, Glass sends an e-mail message instead.)

Saying "OK Glass, send message to Jane Doe" opens a screen where you can speak the message that you want to send to Ms. Doe. We discuss sending instant messages from Glass in Chapter 10.

✔ **"Make a call to. . .":** This command begins a call to the phone number that's associated with the name of your contact. Saying "OK Glass, make a call to John Smith" calls the phone number that's contained within John Smith's contact record. For more about adding and managing your contacts, see Chapter 5; for details on making phone calls with Glass, see Chapter 10.

✔ **"Make a video call to. . .":** This command lets you share what you're seeing or doing with a contact who has a Google+ account. For more about making video calls, see Chapter 10.

We cover voice commands in greater detail in Chapter 8.

Voice recognition on Glass isn't perfect (yet), so if your Glass can't recognize what you're saying, or if you'd rather not tilt your head up and down and speak the commands you see on the menu, tap the Home screen; swipe forward and backward through the available menu options; and select an option by tapping the touchpad.

Touring the Timeline

When you wake up your Glass, you see the timeline screen. Notifications on this screen appear as rows of cards. *Cards* can contain text such as e-mail and instant messages, images, and/or videos. Figure 4-5 shows examples of cards on the timeline screen.

Figure 4-5:
The timeline
screen.

To the left of the Home screen (shown at center in Figure 4-5), you see an area including cards that contain current or future information, such as upcoming events on your calendar, as well as any pinned cards that you want to access quickly. (We discuss pinning cards later in this chapter.) To the right of the Home screen are cards that contain notifications you received in the past, such as e-mail messages.

When Glass adds a new card, that card appears as the first card on the Home screen; older cards move one place to the right. You can move through the row of cards by dragging your finger on the touchpad. You can view more information within a displayed card by tapping the touchpad.

Playing with Cards

You can move between cards on the timeline by using the touchpad and swiping forward and backward, as we discuss in the following sections.

Viewing current and upcoming events

View current and upcoming events in your calendar by swiping backward on the touchpad. These events appear on the timeline (refer to Figure 4-5). Figure 4-6 shows an example information card for an upcoming flight.

Figure 4-6:
An
upcoming-
flight card.

Viewing recent events

Scroll through all timeline cards on the screen by swiping forward or backward on the touchpad. Each card displays information about recent events, such as pictures taken, videos recorded, and messages sent — in other words, every activity that you've performed recently.

Figure 4-7 shows a card displaying the results of a recent Google search on the keywords *time in San Francisco.* You can find out more about doing a Google search in Chapter 9.

Figure 4-7:
Results of
a Google
search.

time in San
Francisco

12:43pm
Time in San Francisco,
CA, USA

.just now

Choosing commands for cards

When you tap any card on the timeline, a menu appears, displaying all the actions you can perform (see Figure 4-8). You can move between menu options by swiping forward and backward on the touchpad and then select an option by tapping the touchpad.

Figure 4-8:
An example
menu option
for a card.

↱ Send   ↱ Share   🗑 Delete

Menu options vary from card to card, and some cards may not have any menu options.

Here are some of the most common menu options for cards:

- ✔ **Share:** This command shares your picture or video with your contacts, your circles in Google+, and other social networks (such as Facebook) you're using with Glass. Find out more about sharing pictures and videos in Chapter 8.

- ✔ **Pin:** This command pins the selected card to the timeline. We discuss pinning cards later in this chapter.

- ✔ **Reply:** This command replies to the selected message by transcribing your spoken message into text.

- ✔ **Delete:** The Delete command removes the selected card from the timeline.

- ✔ **Call:** This command is available only if your Glass is connected to your Android smartphone through a Bluetooth connection so that you can make phone calls.

- ✔ **Read Aloud:** When you choose this command, Glass reads the text on the card so that you can listen to its content on the speaker.

If you'd rather go back to the card without making any menu choices, close the menu by swiping down on the touchpad.

Working with bundled cards

Some cards have a folded top-right corner, which indicates that the card is bundled. A *bundled card* is a group of cards containing information that belongs together. A bundle might contain a card with the current temperature and another card with the extended forecast, for example (see Figure 4-9).

Figure 4-9: A bundled weather card.

When you tap a bundled card, you see the cards within the bundle. You can move between individual cards within the bundle by swiping forward or backward.

Receiving notifications

If your Glass is connected to the Internet through a Wi-Fi connection, it continually keeps track of your online activities (such as your e-mail account), and if something new happens, Glass adds a notification card to the timeline immediately.

When you receive a new notification, you hear a chime from the Glass speaker. You also see the new notification card, with the text `Just Now` in the bottom-right corner, as shown in Figure 4-10. This new card is displayed for five seconds after you hear the chime. When that period elapses, the notification appears on the timeline.

Figure 4-10:
A new
notification
card.

Within that five-second window, you can view the entire notification card onscreen by tapping the touchpad or by tilting your head up. A notification card may also include menu options, such as one to delete the notification; you can access the menu while you're viewing the entire card by tapping the touchpad.

To see missed notifications, swipe forward on the touchpad to see your recent notifications in the order in which they arrived.

Pinning cards

You can pin a card to the Information area easily by tapping the card and then tapping Pin on the resulting menu, as shown in Figure 4-11. To unpin a card and remove it from the Information area, tap it and then tap Unpin on the resulting menu.

Figure 4-11:
The menu option for pinning a card.

Chapter 5

Enhancing Glass with the MyGlass Website

Each day I live in a glass room unless I break it with the thrusting of my senses and pass through the splintered walls to the great landscape.

— Mervyn Peake

*I*n this chapter, we introduce the MyGlass website, which you can access from your computer, tablet, or smartphone. MyGlass allows you to tweak your Glass settings so that the device works best for you.

If you want to add your Glass to another Wi-Fi network, such as a network offered by a hotel where you're staying, you can do this with MyGlass. You can also locate a misplaced Glass with MyGlass so that you can get your device back in your hands as soon as possible.

In addition, you can use MyGlass to enable or disable *Glassware,* which is a Google term for web-based apps that can send data to and receive data from Glass, as well as apps that are written specifically to run on Glass. This enabling and disabling feature allows you to hide Glassware so that your Glass screen shows only the apps you want to use.

Taking the Red Carpet Tour of MyGlass

To tour MyGlass for the first time, fire up your favorite browser on your computer, tablet, or smartphone and then follow these steps:

1. **In the browser address bar, type** `www.google.com/myglass` **and then tap or press Enter.**

   If you aren't logged in to your Google account already, you see the Google login page so that you can log in to your Google account.

   If you're already logged in to Google, you may see a warning page similar to Figure 5-1. Don't panic; this warning screen is a good thing. Google is making sure that you are who you say you are, because it doesn't want an unauthorized user changing your Glass experience.

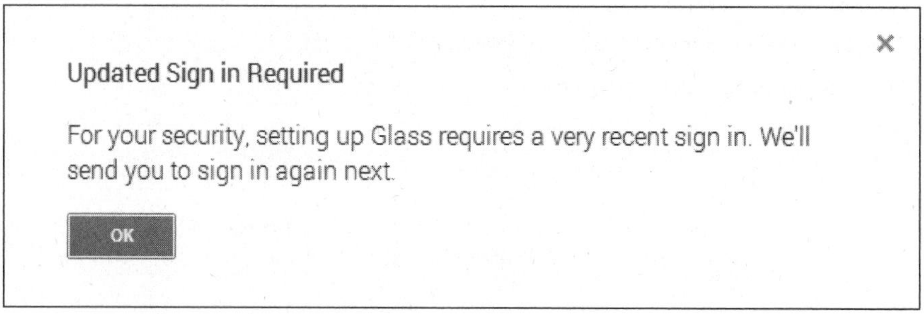

Figure 5-1:
The
Updated
Sign In
Required
warning
page.

✕

Updated Sign in Required

For your security, setting up Glass requires a very recent sign in. We'll send you to sign in again next.

2. **Tap or click the OK button and then type your Google e-mail address and password in the Accounts screen.**

   After you log in, the MyGlass home page appears, as shown in Figure 5-2. The home-page content is organized in a row of cards, just as in your Glass timeline, so you'll feel right at home. A menu bar appears at the top of the home page, with menu options at the left end and your Google e-mail address and account photo at the right end.

3. **Do any of the following:**

   • Select Setup if you want to set up Glass for the first time or after you perform a factory reset.

   • Select MyGlass or the Glass logo to return to the MyGlass home page.

   • Select Glassware so you can shop for Glassware, add Glassware to your Glass, or remove Glassware from your Glass.

- Select Explorers to visit the Glass Community website at `https://www.glass-community.com`.

- Select Support to open a new tab in your browser and view the Glass support page.

- Select your Google e-mail address or account photo to view your Google account information, add an account, or sign out of MyGlass.

Managing Your Contact List

One of the cards on the MyGlass home page lets you add a contact (refer to Figure 5-2), because Google recognizes that you'll often want to contact your friends via voice call, video call, or instant message. Google also knows that you'll likely prefer telling your Glass to call people to fumbling for your smartphone.

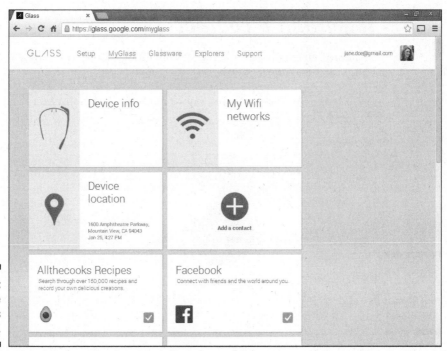

Figure 5-2:
The
MyGlass
home page.

If you don't have a Google Contacts database, you can set one up for free at www.google.com/contacts. You can access your contacts list from your favorite web browser, as well as synchronize your contacts with other programs and devices.

To make your life even easier, Google allows you to import contacts from your Google Contacts database to your Glass by using the MyGlass website or the MyGlass app on your smartphone. When you initiate a voice call, video call, or instant message, your Glass shows a list of your contacts so you can scroll to and select the desired contact.

You can access your imported contacts from the Google Contacts database only by using the touchpad; you can't use the voice feature. That is, when you send a message, you need to scroll through the contact list by swiping backward and forward on the touchpad. When you find the contact you want, tap the touchpad.

Unfortunately, at this writing, MyGlass can store no more than ten contacts that you add within Glass. If your MyGlass contact database is full and you want to add another contact, you have to remove an existing contact first.

Importing a contact

To import an existing contact from your Google Contacts database to MyGlass, follow these steps:

1. **On the MyGlass website, tap or click the Add a Contact card (see Figure 5-3).**

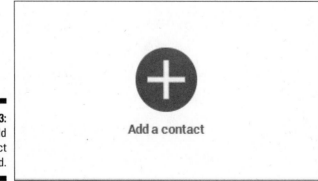

Figure 5-3:
The Add a Contact card.

Add a contact

2. Type the contact name in the Add Contact field, shown in Figure 5-4.

As you type, contact names appear in a list below the Add Contact field. These names are those in your Google Contacts database that most closely match the name you're typing. If you keep typing, the list may grow shorter as MyGlass finds database names that even more closely match the contact you're searching for.

Figure 5-4:
The Add
Contact
field.

3. When you find the name that you want to add (see Figure 5-5), tap or click it.

The new contact appears as a card on the MyGlass home page. You find out more about contact cards later in this chapter.

Figure 5-5:
Add a
contact by
tapping the
contact's
name.

Adding a contact

If you want to add a contact, here's how to do it:

1. **On the MyGlass website, tap or click the Add a Contact card.**

2. **Tap or click the Create New Contact link (refer to Figure 5-4).**

   The New Contact page appears as a new tab in your web browser (see Figure 5-6). If you're using the MyGlass app on your Android smartphone, your default contact management app appears on your smartphone's screen.

3. **Add contact information, including the contact's name, photo (if any), e-mail address, phone number, and birthday.**

4. **Tap or click the Save Now button in the top-right corner.**

5. **Tap or click the MyGlass tab in your browser, or switch to the MyGlass app on your smartphone.**

6. **On your Glass, import the contact as we describe in "Importing a contact" earlier in this chapter.**

If you don't see your new contact, refresh the browser window by pressing F5 on your computer keyboard; on your smartphone, tap Refresh on the menu.

Figure 5-6:
The New
Contact
page.

Viewing a contact card

A contact card (see Figure 5-7) displays the same information on the MyGlass website and in the MyGlass app: the contact's name, a photo of the contact (if any), and icons in the top-left corner that show what type of information the card contains.

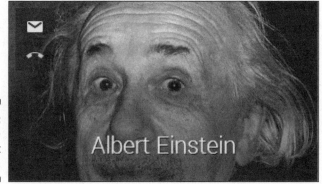

Figure 5-7:
Example
contact
card.

In Figure 5-7, the envelope icon indicates that the contact has an e-mail address, and the telephone-receiver icon shows that the contact has a phone number.

You can get more information about the contact by tapping the contact card. If the contact has more than one e-mail address and/or more than one phone number, a blue check box appears to the right of the contact's preferred e-mail address and/or phone number, as shown in Figure 5-8.

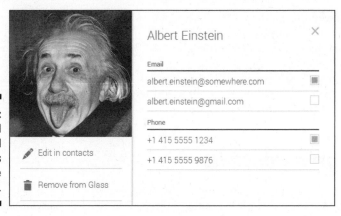

Figure 5-8:
Preferred
e-mail
address
and phone
number.

If you want to set a new preferred e-mail address and/or phone number for a contact, tap the check box to the right of that address or number (refer to Figure 5-8). The check box turns blue to indicate that the address or number is now the preferred one.

When you finish viewing a contact card, close it by tapping the Close icon (the X) in the top-right corner.

Editing a contact

MyGlass makes it easy for you to edit a contact. Follow these steps:

1. **Tap the contact card on the MyGlass home page.**
2. **Tap or click Edit in Contacts below the contact photo.**

    The Google Contacts website opens.
3. **Edit the contact information.**
4. **Save your changes by tapping or clicking the Save Now button in the top-right corner of the page.**
5. **Click the MyGlass tab in your browser or switch back to the MyGlass app on your Android smartphone.**

    Your changes appear in the contact tile.
6. **Close the card and return to the MyGlass home page by tapping the Close icon in the top-right corner.**

Removing a contact

You can remove a contact from MyGlass by clicking or tapping the contact card in the MyGlass screen and then clicking or tapping Remove from Glass (refer to Figure 5-8 earlier in this chapter). MyGlass removes the contact from its database.

Removing a contact removes it only from MyGlass. The contact remains safely stored in Google Contacts.

Merging contacts

Synchronizing your Google Contacts database with Glass can take a while if your database has multiple contacts for several people. Fortunately, Google Contacts makes it easy to clean up your database by merging information in all contacts that have the same name. To merge contacts, follow these steps:

1. **Open the Google Contacts webpage at** `https://www.google.com/contacts`.

2. **Log in to Google Contacts, if necessary, using your Google e-mail address and password.**

3. **In the contact list, tap or click the More button and then tap or click Find & Merge Duplicates on the resulting menu (see Figure 5-9).**

   You see a list of contacts that contain duplicate data, such as the same e-mail address.

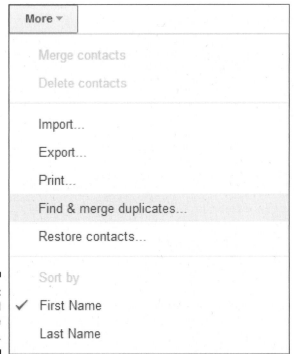

Figure 5-9: Select Find & Merge Duplicates.

4. **Merge the contacts by tapping or clicking Merge Contacts at the top of the list.**

You can view information about merging two or more contacts manually at `https://support.google.com/mail/answer/165334?hl=en`.

Adding a Wi-Fi Network by Using the MyGlass Web App

When you're out and about, you won't have access to the Wi-Fi network that you use on a regular basis in your home or office. If you're going to be using a Wi-Fi network in your office, at a hotel, or even at your favorite coffee shop, you can add a new Wi-Fi network for your current location within the MyGlass website (www.google.com/myglass) and save it the next time you're there.

You'll find two types of Wi-Fi networks:

- **Open:** An *open* network allows anyone to use your Wi-Fi connection because it isn't secured with a password.
- **Protected:** A *protected* network requires authentication to gain access to network resources. To use and connect to a protected Wi-Fi network, you need three things: a network name, an encryption method, and the network password.

MyGlass makes it easy for you to set up a new Wi-Fi network that you can store on your Glass so you can connect to the network easily when you're in range of that network.

Generating a QR code

The first step in connecting to a Wi-Fi network in MyGlass is generating a QR (Quick Response) code for that network. You can create a QR code for two types of Wi-Fi networks: an open network, which doesn't require a password, and an encrypted (or hidden) network, which requires a password. (For more on QR codes, see Chapter 3.)

For an open network

If you're at a location that offers public Wi-Fi, such as a coffee shop or an airport, here's how to create a QR code for that public network:

1. **Tap or click the My Wifi Networks card on the home page (see Figure 5-10).**

   The Add Network card appears so that you can add information about your Wi-Fi network.

Figure 5-10:
The My Wifi
Networks
card.

2. **Type your wireless network name in the Enter Network Name field.**

3. **Type the password (if one is required) in the Click to Enter Password field.**

   Passwords are case-sensitive.

4. **Tap or click Generate Code (see Figure 5-11).**

   This link is active only when you've entered all the information for your Wi-Fi network. If the link is disabled, check your settings to ensure that you named your network properly, selected the correct encryption method, and entered a password if necessary.

   After you select Generate Code, a QR code appears onscreen. This code allows you to add the Wi-Fi network, which we discuss in "Adding the network to your Glass" later in this chapter.

Figure 5-11:
Generating
a QR code
for an open
network.

For a hidden network

To create a QR code for a hidden network, follow these steps:

1. **Tap or click the My Wifi Networks card on the home page (refer to Figure 5-10).**

   The Add Network card appears (refer to Figure 5-11).

2. **At the bottom of the card, tap or click the Encryption Type text link.**

3. **Type your wireless network name in the MyWifiNetwork field.**

4. **Type the password in the password field.**

   Note that passwords are case-sensitive.

5. **Select the Wi-Fi encryption method by tapping or clicking the WPA/ WPA2 button.**

   You see three encryption options in the drop-down menu: None, WPA/ WPA2, and WEB (see Figure 5-12).

Add Network

Network name and password are case sensitive.

MyWifiNetwork

••••••

WPA/WPA2 ▾

Generate Code

Figure 5-12:
Choose an
encryption
option.

6. **Select the appropriate encryption method.**

   If you don't know which encryption method the network uses and/or the network password, consult the staff of the location that offers the Wi-Fi service.

7. **Tap or click Generate Code.**

   This link is active only when you've entered all the information for your Wi-Fi network. If the link is disabled, check your settings to ensure that you named your network properly, selected the correct encryption method, and entered a password (if necessary).

   After you select Generate Code, a QR code appears onscreen, as shown in Figure 5-13. This code allows you to add the Wi-Fi network, which we discuss in the next section.

Figure 5-13:
The generated QR code for a hidden network.

Adding the network to your Glass

Here's how to add a Wi-Fi network to your Glass with a generated QR code (refer to the preceding section):

1. **Swipe backward on the touchpad until you see the Settings bundle card.**

2. **Tap the Settings bundle card.**

   The Settings screen appears.

3. **Tap the Wi-Fi Settings card (see Figure 5-14).**

Figure 5-14:
The Wi-Fi Settings card.

4. Tap the resulting Join Network menu option (see Figure 5-15).

Figure 5-15:
The Join
Network
menu
option.

5. Tap the resulting Add Wifi Network card (see Figure 5-16).

If Glass detects Wi-Fi networks available near you, you see one of the network cards onscreen. Scroll through all the networks by swiping forward on the touchpad. After you scroll past the last network card, you see the Add Wifi Network card.

Figure 5-16:
The Add
Wifi
Network
card.

6. When your Glass prompts you to point to the QR code on your screen (see Figure 5-17), align the box on your Glass display with the QR code on your screen.

When Glass recognizes the QR code, it connects to the Wi-Fi network automatically and chimes when the connection is complete.

Figure 5-17:
Glass
prompts
you to point
to the QR
code.

Point towards the QR code on your screen

7. **On the MyGlass site, close the card with the QR code by tapping or clicking the Close icon (the X) in the top-right corner.**

   You return to the MyGlass home page.

   On your Glass, you see your new Wi-Fi network connection onscreen (see Figure 5-18).

NewWifiNetwork
Connected

WPA

Figure 5-18:
The
NewWifi
Network
Connected
card on
Glass.

8. **Return to the Home screen on your Glass by swiping down on the touchpad.**

Tracking Down Your Glass

The MyGlass app for your Android smartphone includes the nifty Device Location card, which helps you locate your Glass as quickly as possible in case you misplace it.

The Device Location card works if your Glass has a Wi-Fi connection or is paired with a smartphone that has the MyGlass app installed on it. If you haven't done so yet, find out how to pair your Glass and your smartphone in Chapter 3.

The Device Location card in MyGlass (see Figure 5-19) shows you the current location of your Glass. It also shows the last time MyGlass checked your smartphone to get the location of your Glass.

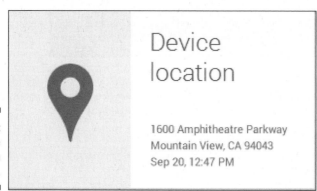

Figure 5-19:
The Device
Location
card.

To view a map showing the device's current location, click or tap the card. A map displays the pinned location of your Glass device, as shown in Figure 5-20. Unfortunately, this display is as close as MyGlass can get to finding your Glass, but it should give you enough information to search the area or call the authorities.

Figure 5-20:
A map
displays
the current
location of
your Glass.

You can view the location in Google Maps by tapping or clicking the View on Google Maps link.

Enabling or Disabling Glassware

Glassware apps can send content to and receive content from your Glass. Some Glassware apps are preinstalled. You can use MyGlass to shop for more Glassware and to customize the apps on your Glass by tapping the Glassware menu option at the top of the MyGlass home page.

See Chapter 12 for a complete discussion of preinstalled and third-party Glassware apps.

A list of Glassware apps appears on the MyGlass home page. Each Glassware app card contains the name of the app, a brief description, and an On/Off slider button (see Figure 5-21).

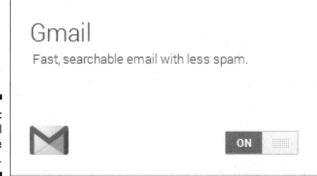

Figure 5-21: The Gmail Glassware app card.

If the slider displays the blue On button, the Glassware app is enabled. You can disable the app by tapping or clicking the gray Off button.

To view more information about a Glassware app, tap or click its app card. A new window opens, displaying detailed information about the app. In the bottom-left corner of the window, you also see the permissions and Glass resources that the app requires.

Chapter 6

Managing Glass Settings

· ·

· ·

I love inventing worlds and characters and settings and scenarios.

— Jerry B. Jenkins

*G*lass is a remarkably flexible system that allows you to obtain and change information and connections to fit your needs. You may find yourself at a coffee shop that offers a free Wi-Fi connection (and what self-respecting coffee shop doesn't?), but you need to set up a new connection on your Glass. What do you do?

It's easy to access a list of Settings cards from the Home screen and then set up your Wi-Fi connection. But setting up Wi-Fi connections is just one of the things you can do on Glass. You can also

✔ Set up a connection with a Bluetooth device such as your smartphone

✔ Tell Glass how it should wake up when you tilt your head

✔ Get information about the device, such as how much memory is available for storing photos, videos, and timeline items

✔ Change the volume settings of the speaker so your ear and brain are comfortable

In case Glass is having trouble operating correctly and you suspect that an app is to blame, you don't break Glass the way you break glass when you want to get a fire extinguisher or fire hose. Instead, you can reset your Glass to its original factory settings — the ones it had when you first took it out of the box. Then you can start running Glass again and see whether the trouble continues.

Finding the Settings Bundle

You can find the Settings bundle by swiping backward on the touchpad until you see the bundle, as shown in Figure 6-1.

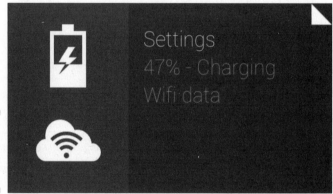

Figure 6-1:
A Glass
Settings
card bundle.

You can tell that you're looking at a bundle because the top-right corner of the card contains a white right triangle and the card itself displays the battery status.

When you tap the Settings card bundle, you see several individual settings cards:

- ✔ **Wifi:** The Wifi Settings card shows the status of your Wi-Fi network connection. (See "Doing the Old Join and Switch" later in this chapter.)

- ✔ **Bluetooth:** The Bluetooth Settings card shows the status of the active connection with a Bluetooth device. (See "Pairing with Bluetooth Devices" later in this chapter.)

- ✔ **Device Info:** The Device Info Settings card gives you vital information about your Glass, including storage and operating system. (See "Seeing What's Going On" later in this chapter.)

✔ **Head Wake Up:** The Head Wake Up Settings card allows you to set the angle at which you need to tilt your head to turn on the screen. (See "Using Settings That Use Your Head" later in this chapter.)

✔ **On-Head Detection:** The On-Head Detection Settings card turns Glass on as soon as you put it on your head. (See "Using Settings That Use Your Head" later in this chapter.)

✔ **Screen Lock:** The Screen Lock Settings card allows you to set a pattern-based password and lock the screen whenever you take Glass off your head or deactivate the device. When you put Glass back on your head or activate it, you need to enter the pattern on your touchpad to access screens and functions. (For details on setting Screen Lock, see this book's Web Extras at `www.dummies.com/extras/googleglass`.)

✔ **Wink:** The Wink Settings card allows you to activate the Wink feature, which lets you take a photo by winking your right eye — even if the Glass display is off. (To find out more about setting up Wink, see Chapter 8.)

✔ **Volume:** The Volume card shows the current volume level of your speaker as a percentage. (See "Getting the Volume Just Right" later in this chapter.)

We cover battery charge in Chapter 2. For details about tweaking your Glass by using settings cards, read on.

Seeing What's Going On

When you're in the Glass Settings screen, you can get more information about the device by tapping the Settings card bundle and then tapping the Device Info card, shown in Figure 6-2.

Device info
XE12 up to date
12.1 GB free

Serial XXXXYYYYZZZZ

Figure 6-2:
The Device
Info card.

Storage

Glass contains 16GB of storage space, 12GB of which is allocated for pictures, video, and other timeline attachments. If you have a Google+ account, your photos and video are automatically synced with your Google+ Auto Backup folder (see Chapter 8).

You can get more information about the Google+ Auto Backup folder, including how to turn the folder on and off and how many files you can back up to the folder, at `https://support.google.com/plus/answer/1647509?hl=en`.

Operating system

The bottom of the Device Info card shows the current version of the Glass operating system that's running. It also indicates whether an operating-system update is available. The next time you plug Glass into the charger, it downloads and installs any operating-system update, even if you don't check to see whether an update is available.

All Glass updates are downloaded through your Wi-Fi connection automatically, but the device has to be connected to the Internet for that to happen.

Getting the Volume Just Right

By default, speaker volume on your Glass is set to 100 percent. Because Glass makes sounds the first time you turn it on, you'll know whether this speaker setting is too loud or just right.

Changing the default volume setting is easy. Just tap the Settings card bundle; tap the Volume Settings card, shown in Figure 6-3; and then slide your finger forward and backward on the touchpad to move the onscreen volume slider.

When you release your finger, the slider disappears, and the speaker chimes so that you can determine whether the new volume is right. If you need to adjust the volume more, tap the Volume Settings card again and reset the slider.

Figure 6-3:
The Volume
Settings
card.

Doing the Old Join and Switch

Many coffee shops, restaurants, and other businesses have free Wi-Fi networks available to persuade customers to stay a while, perhaps buy more products or services, and attract other customers to meet you at that location (and, of course, buy things themselves).

Glass automatically detects Wi-Fi networks in your immediate area so that you can join one of them. Then you can browse the Internet, check your e-mail, and browse your social networking profiles without having to look at your smartphone screen or bring a laptop with you.

Joining a network

If you're at a location that offers public Wi-Fi service, such as at a coffee shop or airport, your Glass finds all public Wi-Fi connections near you. These connections don't require a password for you to access their networks.

Here's how to join a public Wi-Fi connection that your Glass finds:

1. **Swipe backward on the touchpad until you see the Settings card bundle (refer to Figure 6-1).**

2. **Tap the Settings card bundle.**

   The Wifi Settings card appears, as shown in Figure 6-4.

Figure 6-4:
The Wifi
Settings
card.

3. **Tap the Wifi Settings card to view all available Wi-Fi network connections near you.**

4. **Swipe forward on the touchpad until you see the network you want to join.**

5. **Tap the touchpad and wait for Glass to connect you to the network.**

Another way to join a Wi-Fi network is to use a QR code in MyGlass — a requirement if you want to join a secured Wi-Fi network that requires a password. See Chapter 5 for details.

Switching networks

When you leave your favorite Wi-Fi hotspot to go to another location, such as your home network, it's easy to switch to that network on your Glass. Follow these steps:

1. **Tap the Settings card bundle.**

2. **Tap the Wifi Settings card.**

3. **Tap Switch Network in the resulting menu.**

   You see a list of the networks that Glass has detected.

4. **Tap the name of the network you want to join.**

   Glass automatically disconnects you from the active open network and switches you to your selected network.

If you've already added details about a Wi-Fi network on the MyGlass website or on the MyGlass app for smartphones, you can switch to that Wi-Fi network without having to set up any other information. If you don't have information about the Wi-Fi network that Glass found, you need to set up the Wi-Fi connection.

Disconnecting from a network

What do you do if you don't want to be connected to a wireless network anymore? Tell Glass to forget the network by tapping the network's name in the Wi-Fi connection card and then tapping Forget on the resulting menu. When you forget a network, Glass disconnects you from the Wi-Fi connection. It also removes this connection from memory, so if you want to connect to the network again later, you have to join the network again.

Pairing with Bluetooth Devices

To check the current Bluetooth status of your Glass, tap the Settings bundle card and then tap the Bluetooth Settings card, shown in Figure 6-5.

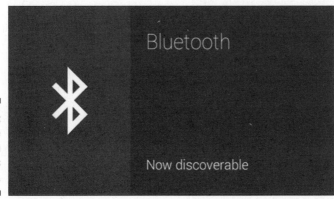

Figure 6-5:
The Bluetooth Settings card.

Bluetooth connections require each device to be discoverable. In the case of Google Glass, you have to tell your smartphone that you want to pair it with your Glass. To see how this process works, connect *(pair)* your smartphone and your Glass via Bluetooth by following these steps:

1. **Open the Bluetooth Settings screen on your smartphone.**

2. **On your Glass, tap the Settings card bundle and then tap the Bluetooth Settings card.**

3. **On your smartphone, select Search for Devices.**

4. **When the smartphone finds the device, tap Glass Device in the list of available devices.**

   On your smartphone's screen, you see the Bluetooth Pairing Request window (see Figure 6-6), which includes a passkey. On the screen of your Glass, you see a card with the same passkey.

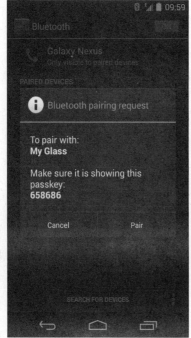

Figure 6-6:
The
Bluetooth
Pairing
Request
window.

5. **Make sure that the passkeys on your Glass and your smartphone match.**

If the passkeys don't match, you should contact your smartphone manufacturer and/or a Glass Guide for more assistance. You can find out more about Glass Guides and getting help in Chapter 17.

6. **Pair the devices by tapping the Pair button on your smartphone and then tapping the touchpad on your Glass.**

   Glass displays a card stating that pairing was successful. After a few seconds, Glass returns you to the Bluetooth Settings card, which looks similar to Figure 6-7.

Figure 6-7: The Bluetooth Settings card displays a successful connection.

Nexus 5
Headset
Data from Wifi
MyGlass

Now discoverable

More Bluetooth fun

Even if your Glass isn't paired with a Bluetooth device, the Bluetooth Settings card lets you do three things:

✔ **Use your Glass as a Bluetooth headset.** If you want to use your Glass as a headset to talk on the phone with friends, you can do that by setting up a Bluetooth connection between your Glass and your smartphone.

✔ **Use a Bluetooth data connection.** Glass constantly searches for Bluetooth devices to which it can connect. If you've paired

Glass with your smartphone, you can connect to the Internet on your Glass by using your smartphone's data connection.

✔ **Check the status of your MyGlass app.** If you have the MyGlass app installed, the Bluetooth Settings card shows whether your connection to the app is active.

Using Settings That Use Your Head

You have three ways to turn on the display:

- ✔ Tap the touchpad.
- ✔ Use the Head Wake Up feature.
- ✔ Use On-Head Detection.

In this section, we discuss how to configure the ones that use your head.

Configuring Head Wake Up

Head Wake Up lets you turn on the display by tilting your head up instead of tapping the touchpad. To access this feature's settings, tap the Settings card bundle and then tap the Head Wake Up Settings card (see Figure 6-8). This card shows whether the feature is enabled and also shows its *wake angle* — the angle at which you need to tilt your head to wake the device.

Figure 6-8: Head Wake Up Settings card.

By default, the wake angle is 30 degrees. Fortunately, Google recognizes that you may not find this setting comfortable, especially if you tilt your head often. You can change the default wake angle, as follows:

1. **Tap the Settings card bundle and then tap the Head Wake Up Settings card.**

2. **When the menu appears, swipe forward to find and tap Set Wake Angle.**

   The screen shown in Figure 6-9 appears.

Figure 6-9:
Change
the wake
angle in this
screen.

Wake angle is

30°

Adjust head and tap to set

3. **Tilt your head to the angle you'd like to use to wake the display.**

   The angle size appears in degrees onscreen as you move your head.

4. **When you feel comfortable with the angle, set it by tapping the touchpad.**

Configuring On-Head Detection

Instead of having to remember to press the power button to turn Glass on or off, you can enable On-Head Detection instead. This feature activates Glass when it's on your head and deactivates it when the device isn't on your head. When Glass is off, the device is still powered on, but its touchpad, display, and audio are off, and it won't pick up any incoming phone calls.

To configure the On-Head Detection feature, follow these steps:

1. **Tap the Settings card bundle and then tap the On-Head Detection Settings card (see Figure 6-10).**

2. **Tap Calibrate on the resulting menu.**

3. **When your Glass prompts you to do so, take it off for a few seconds.**

   As you take the device off, the speaker plays a descending chime.

4. **Put Glass back on your head and then tap the touchpad.**

   When you put the device on your head again, the speaker plays a rising chime, and you can interact with your Glass again.

If you find that On-Head Detection isn't working as well as you'd like, tap the On-Head Detection Settings card to turn off this feature. Then recalibrate On-Head Detection by following the steps in this section.

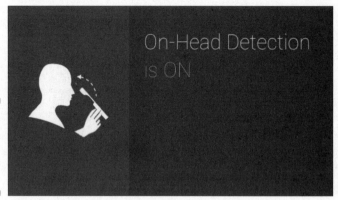

Figure 6-10:
On-Head
Detection
Settings
card.

Resetting Glass

Glass is a computer, and like all computers, it may not always run as well as it should, or an app may not function properly. In such cases, you may want to reset your Glass. When you reset Glass, all data on it is deleted, and all the settings that came on it when it left the factory are restored, so the device is just like new.

Be careful when you reset Glass, because when you do, you'll lose that data forever — with the exception of any data that you downloaded to Glass from your Google+ Auto Backup folder (see "Storage" earlier in this chapter). As you do with any computer, make sure that the important data on your Glass is stored elsewhere (such as your Google+ Auto Backup folder) before you reset the device.

If you need to reset Glass to its original factory settings, tap the Settings card bundle, tap the Device Info Settings card, and then choose Factory Reset from the menu shown in Figure 6-11.

Figure 6-11:
The Factory
Reset option
has a poison
icon to
underscore
the fact that
all data on
your Glass
will be
deleted.

Glass asks you to confirm that you want to reset it, as shown in Figure 6-12. Begin the reset process by tapping the touchpad. If you want to abort the process and return to the Device Info Settings card, swipe down on the touchpad.

Figure 6-12: You have one last chance to abort the factory-reset process.

You'll know that the reset process is complete when your Glass displays the setup screen. If you want more information about how to set up Glass, turn to Chapter 3.

Chapter 7

Using Glass Responsibly

> *You must take personal responsibility. You cannot change the circumstances, the seasons, or the wind, but you can change yourself. That is something you have charge of.*
>
> — Jim Rohn

When Google Glass was introduced in spring 2013, numerous web articles appeared to tell you — and, presumably, the Google Glass Explorers who were testing it — all about Google Glass etiquette. In your favorite search engine, just search for *google glass etiquette* to see the list for yourself.

But there's no need to stress; getting inside information like this is why you're reading this book. In this chapter, we list some of the ways you should and shouldn't use your Glass.

For more advice about what *not* to do, see Chapter 19.

There's a Computer on Your Head

Wearing a Glass in everyday life can create some unintended consequences. One shining example occurred in late October 2013, when a driver in California was cited for wearing her Google Glass while driving (see Figure 7-1). According to the citation, she was violating a state law that requires people not to drive a vehicle equipped with a monitor unless it displays "global mapping displays, external media player (MP3), or satellite radio information." The driver fought the ticket, however, because she didn't have her Glass turned on at the time, and the officer couldn't prove otherwise.

Figure 7-1:
Driving with
Glass.

In early 2014, a judge in traffic court agreed with the driver. The charges were thrown out because there was no evidence beyond a reasonable doubt that her Glass was turned on at the time. Though this episode had a happy ending, it also illustrates the brave new world that we're entering with wearable computers and acts as a reminder that (as with any piece of technology) a set of unwritten etiquette rules comes with your Glass.

There's a Camera on Your Head

Sometimes, it's hard to remember that you have your Glass on your head because it's so light and unobtrusive, which may be why the driver who was cited for wearing her Glass (see the preceding section) didn't even think about removing it. The people you interact with, however, may behave a bit oddly toward you. That behavior may be caused by your inability to maintain eye contact with them because you're too busy looking at the screen or moving your head up to activate your Glass.

Also, people may not be sure that you're not taking photos or recording video surreptitiously. Experienced users can tell when you're taking a photo or video (the Glass screen lights up, and you have to speak a command or push a button), but not everyone is familiar enough with this new technology to recognize these signs.

When you're talking with people, it's a good idea to take your Glass off your head. That way, they'll know that you're giving them your full attention and that you're not secretly recording the conversation.

Ask permission before recording

You can't control what photos and videos other people take on Glass, and the general rule applies that a few bad apples can spoil things for everyone else. By the time you read this book, those bad apples may have already made it difficult for everyone else to take photos and videos on Glass without arousing suspicion.

You can go a long way toward alleviating that suspicion by always asking permission before taking photos and videos of anyone — or of any business, or of anyone's pets or children.

Share responsibly

This guideline was illustrated expertly by popular tech blogger Robert Scoble, an early tester and proponent of Glass who showed his enthusiasm by taking a picture of himself with his Glass on his head. . .as he was taking a shower.

Though Scoble was discreet, the photo didn't exactly kindle enthusiasm among potential users. Instead, people (especially those in the media) took note that Glass doesn't tell people other than the wearer that the camera is on, so it could encourage all sorts of mischief.

Make Miss Manners proud: Immediately share your photos and videos with the person who gave you permission to take them (see the preceding section) and then ask whether it's all right to share them with anyone else. If not, don't post those items on the web. Instead, take a minute to delete the photos from your Glass and any other file-storage services you use, such as your Google+ backup folder.

Don't Be a Showoff

You may have already been annoyed by someone who talks loudly on his smartphone in a public place. You'll be the person people love to hate if you make phone calls on your Glass, start speaking commands to it, or dictate e-mail messages to it in public. Don't show off with Glass, either intentionally or unintentionally.

One way to avoid being a showoff is to avoid wearing your Glass to a party or other social gathering. It's nice to be able to check a person's name and interests before you talk to her, but she may be uncomfortable when you recite all her interests instead of asking her for that information.

Also, don't use Glass to correct someone else's grammar or to try to show that you're superior to them because you have the coolest technology. If you do, don't be surprised to find people avoiding you. (If you're surprised or if you feel the need to feel superior, we can't help you.)

Be a Glass Ambassador

When you wear your Glass, people are going to ask you about the device and what it can do. They may simply be curious, or they may be afraid that you're going to do something nefarious with your Glass.

So be gracious. Answer all questions as completely as you can, and if you can't find the answers, ask the other person for his preferred contact information so you can send him the answers when you find them.

Know That Wearing Glass Is a Privilege

Use Glass wisely, because courtesy matters. If you fret that courtesy isn't part of society anymore, here's your chance to be a rebel.

If you want to see what could happen if you don't use your Glass wisely, the Mashable website offers a helpful, amusing video at `http://mashable.com/2013/05/16/google-glass-glasshole`. A bonus video on that page shows you how New Yorkers react to a person wearing Glass, which gives you a good idea of the reactions you can expect when you wear your own Glass in public.

Part III
It Can Do That?

6 attractions nearby

🚗 7 min   just now

web extras

Find out how to wink at your Glass to take pictures at www.dummies.com/
extras/googleglass.

In this part . . .

✔ See how to snap photos and shoot videos on Glass.

✔ Browse the web and get the information you're looking for with Google Now.

✔ Use Glass with your smartphone to make and receive calls and to send SMS text messages.

✔ Navigate around the world, using the built-in Glass navigation features.

✔ Find out how to wink at your Glass to take pictures at www. dummies.com/extras/googleglass.

Chapter 8

Making Your Memories

> *What I like about photographs is that they capture a moment that's gone forever, impossible to reproduce.*
>
> — Karl Lagerfeld

Google Glass was made for taking pictures and videos. The device is equipped with a 5-megapixel camera and lots of operating-system bells and whistles designed to give you the best results from your pictures or videos. The camera, for example, uses high dynamic range (HDR) imaging technology to detect low-light environments and adjusts automatically to capture brighter, sharper pictures even when the subjects are moving.

Instead of fumbling for your smartphone and then launching a camera app, you can take pictures and videos with your Glass by speaking a few words or even blinking your eyes — especially convenient when you can't use your hands, such as while riding your bike.

The real power of capturing media on your Glass, however, is in using the Google+ service to add captions to your pictures and videos, share them with friends, and back up your files from your Android or iOS smartphone. We show you how to do all that in this chapter.

Snapping Pictures

When Glass takes a picture, it captures an image based on what you're currently looking at. Glass gives you three ways to take a picture:

✔ At the Home screen, say "OK Glass, take a picture."

✔ Press the Camera button in the top-right corner of the Glass frame, above your right eye. (You can use this button to take a picture even when the Glass display is off.)

✔ When the Wink feature is on, you can take a picture by winking your right eye.

✔ At the Home screen, tap the touchpad and then choose Take Picture from the menu shown in Figure 8-1.

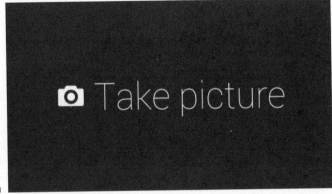

Figure 8-1:
Choose
Take Picture
from this
menu.

After your Glass takes the picture, you see a preview onscreen, as shown in Figure 8-2. You can share the picture from the screen by tapping the touchpad or by saying "OK Glass, share with. . ." followed by the name of the contact or Google+ circle you want to share the picture with. (You can also share it with a Facebook friend if you have the Facebook Glass app installed.) We discuss sharing in more detail later in this chapter.

Figure 8-2:
Glass is
waiting for
voice com-
mands after
taking a
picture.

To view the picture in your timeline, swipe forward on the touchpad at the Home screen. Tap the touchpad again to share or delete the picture from your timeline. You can also find the picture in your Auto Backup album on Google+ after Glass has a chance to sync your pictures (see "Backing Up Your Media to Google+" later in this chapter).

Winking to take a picture

Your Glass detects a wink, and the Wink feature makes good use of this eye-blinking capability so you can take a picture just by winking your right eye.

You should always let people know that you're going to take a picture of them by winking before you do so. Otherwise, those who think you're winking at them may let you know that they don't appreciate it.

Before you can wink to take a picture, you need to activate the Wink feature, like so:

1. **On the timeline screen, tap the Settings card.**

2. **Tap the Wink Settings card.**

3. **When you're prompted to do so, look at the screen and wink twice so that Glass can calibrate the feature correctly.**

Now that you have the Wink feature enabled, you can use it whether the Glass screen is on or off. You can check whether Wink is on by looking at the Wink Settings card (see Figure 8-3). If you don't want to use the Wink feature, deactivate it by following Steps 1 and 2 above and then choosing Disable on the Wink screen.

Figure 8-3:
The Wink Settings card shows when the Wink feature is on.

Evoking a vignette

In Glass parlance, a *vignette* superimposes Glass features on your screen over the picture you've just taken. This feature is a neat little way to share your experience of taking pictures on Glass. Here's how to make a vignette:

1. **Make sure that the screen displays the feature (such as the weather forecast) you want to use.**

2. **Take the picture by pressing the Camera button.**

3. **Tap the touchpad and then choose Make Vignette from the menu.**

   The vignette appears onscreen (see Figure 8-4). If you want to see more menu options (such as one that lets you share the vignette with someone else), say "OK Glass" or tap the touchpad and then choose the desired option from the menu.

Figure 8-4:
A Glass
vignette.

Capturing Videos

Like the cameras on most smartphones sold today, the camera on your Glass allows you to take videos. Capturing videos with your Glass is similar to taking pictures. You have a choice of three ways to start recording:

- At the Home screen, say "OK Glass, record a video."

- Press and hold the Camera button for 1 second. (This button is in the top- right corner of the Glass frame, above your right eye.)

- At the Home screen, tap the touchpad and then choose Record Video from the menu (see Figure 8-5).

While Glass is recording your video, you see that video on your display, as well as a time counter. By default, Glass records video clips for 10 seconds and then stops recording automatically, but you can extend recording time, as we discuss in the next section.

When you finish recording your video, it appears as a card on your timeline, and it also appears in your Auto Backup album on Google+. (For more on Auto Backup, see "Backing Up Your Media to Google+" later in this chapter.)

Extending a recording

The reason for the "10-second rule" for video recording on Glass is that Google thinks you'll want to take brief videos and share them immediately. Recording videos longer than 10 seconds is both possible and easy, though.

Recording requires a lot of processing power. As a result, the longer you record a video, the hotter the Glass device gets. If your Glass gets too hot, it may become too uncomfortable for you to wear and may also start to behave erratically. Keep a close eye (pun unintended) on the temperature of your Glass, and stop recording when it heats up (see "Stopping a recording" later in this chapter).

You have two ways to extend a recording:

- ✔ While the video is recording (see Figure 8-6), press the Camera button.
- ✔ Tap the touchpad and then choose Extend Video from the menu that appears.

The progress bar at the bottom of the display disappears, and Glass continues to record until you tell it to stop (see the next section) or until you run out of battery power or file storage.

Even if you used the "OK Glass" voice command to start recording, you have to use your hands to extend or stop the recording.

Figure 8-6:
Just press Camera to extend the recording.

Stopping a recording

You can stop recording in either of two ways, depending on how you're recording:

- ✔ If you're making a default 10-second recording and want to stop recording before the time limit expires, tap the touchpad, swipe through the video menu, and then choose Stop Recording from the menu.

- ✔ If you're making an extended recording (see the preceding section), press the Camera button once. You can also choose Stop Recording from the menu as you would with a 10-second video.

When you stop recording, the `"ok glass"` prompt appears at the bottom of the screen, as shown in Figure 8-7. If you want to share the video with or send the video to one of your contacts or people in a Google+ circle, say "OK Glass, share with. . ." followed by the name of your contact or circle. We talk about other ways to share later in this chapter.

" ok glass "

Playing video

After you record a video, it appears on your Glass timeline as a card. Tapping the card opens a menu that offers the following choices:

- ✓ **Play:** Plays the video from the beginning (see Figure 8-8).
- ✓ **Share:** Opens another menu that allows you to share the video with one or more contacts, Google+ friends, or various social networks via Glassware apps.
- ✓ **Delete:** Removes the video from your timeline and from your Glass.

▶ Play

You can always pause playback by tapping the touchpad and resume playback by tapping the touchpad again. Also, while the video is playing, you can swipe backward on the touchpad to rewind and swipe forward to skip boring parts.

Capture with care

Taking pictures and videos of yourself or others could create unintended consequences if you're not careful. Plenty of people have been denied jobs or fired for posting inappropriate pictures or videos to their social networking profiles. If you tell your boss that you're sick, for example, and then post a video that you took on your Glass while enjoying a bike ride that same day, he or she may decide that you're not smart enough to continue working for the company.

Also, after you take pictures or videos of yourself, you may want to share them with a trusted friend or two to ensure that the material is appropriate. If you're too embarrassed to show a picture or video of yourself to friends, that's a good sign that you shouldn't show it to anyone else.

Sharing Your Glass Media

Glass is built to share data. You can share pictures and videos from your Glass with your friends, your Google+ circles, and other social media sharing apps available for Glass (such as Facebook and YouTube). This section shows you how.

Google+ keeps an eye out for content in posts that isn't eligible to be shared. You must not share pictures or videos that are owned by another person or by a company without express written permission, for example. If you don't, your post will be removed, and you could be barred from the Google+ service. If you're not certain how you should use your Glass to share information, see Chapter 7.

Sharing from your timeline

It's easy to share a picture or video from your timeline. Just follow these steps:

1. **Find the picture's or video's card by swiping forward and backward on your timeline until the card appears onscreen.**

2. **Tap the touchpad and then choose Share from the menu.**

   You see the Share icon, as shown in Figure 8-9.

3. **Swipe through the available sharing options in the menu and then select the service with which you want to share your picture or video.**

4. **If you want to add a caption, see the next section; otherwise, wait until Glass shares your picture or video.**

   When Glass finishes the sharing process, you see the picture or video onscreen.

Adding captions

Have you ever been frustrated when someone shares a picture with you but doesn't add the filename or a brief description? Posting media without explaining who's in the picture or video and what that person is doing can cause viewers to complain to you — or ignore your posts. Your Glass makes it easy for you to be nice to your viewers by adding captions.

At the bottom of the screen is the *progress bar,* which is a white line. This line moves from left to right and denotes the time it takes to share the picture or video. You can decide whether you want to add a caption while the progress bar is onscreen. When the progress bar reaches the right side of the screen, time runs out; the progress bar disappears; and you're sharing without a caption.

You can add a caption in two ways:

✔ Tap the touchpad after you tell Glass the contact, Google+ circle, or social network (such as Facebook) with which you want to share your picture or video (refer to "Sharing from your timeline" earlier in this chapter).

✔ When you share a picture or video from your timeline, you see `"ok glass"` at the bottom of the screen, as shown in Figure 8-10. This text informs you that you can speak commands, so the other way you can add a caption is to say "OK Glass, add caption."

Figure 8-10:
You can tell Glass to add a caption.

Whichever of the preceding two methods you choose, when the Caption screen appears, say what you want to include in the caption. As you speak, the words appear onscreen so that you can see what the caption will look like (see Figure 8-11). You can include hashtags, if you want; see the nearby side-bar "Cooking up caption #hashtags."

Figure 8-11:
Speak your caption.

If you decide that you don't want to add the caption, swipe down on the touchpad. If you do want to add it, tap the touchpad. Your caption appears in the bottom-left corner of the picture, and text below the picture informs you when the caption was added (see Figure 8-12).

Figure 8-12: Your finished caption, complete with time stamp.

Sharing with your computer

Another way to share your media is to share it with your computer. Just connect your Glass to your computer with a micro USB cable. Your computer recognizes that Glass is a media device, just like a smartphone or camera. Then you can access the Glass files on your computer as you would any other picture and video files.

Cooking up caption #hashtags

If you want other Google+ users to find your pictures based on a specific topic, you can add hashtag topics to your caption. Hashtag topics are preceded by the pound (#) sign, and Google+ searches for them in posts when a Google+ user searches for a topic. By default, Glass adds the #throughglass topic to your caption automatically.

You can add your own hashtag topics to your caption, too. If you say "hashtag biking," for example, Glass automatically converts your spoken text to #biking.

To view your pictures and videos in Windows Explorer, click the Glass device in the folder tree and then navigate to the \DCIM\Camera folder. On a Mac, use the Finder to navigate to the appropriate media folder. Then share, copy, and paste files from that folder as you would from any other folder on your computer.

Backing Up Your Media to Google+

Google automatically synchronizes all your pictures and videos to your Auto Backup album on Google+ when you plug your Glass into the charger and connect it to the Internet via Wi-Fi (see Chapter 3).

The Auto Backup folder is part of the Google Drive account that's part of your Google account. You can find out more about Google Drive at https://support.google.com/drive/answer/2424384?hl=en.

Pictures and videos upload at their original resolutions, up to a maximum 4,288 by 2,848 pixels. Google provides 15GB of free storage space in the Auto Backup folder, which should be enough for most needs.

If the pictures in the Auto Backup folder occupy more than 15GB, Google automatically resizes the pictures to standard size: a maximum 2,048 pixels on the longest edge. This resizing frees some room in the Auto Backup folder for more files.

Google allows you to back up an unlimited number of videos to the Auto Backup folder as long as each video is less than 15 minutes long and has a resolution of 1080p or lower. (1080p resolution is 1,920 by 1,080 pixels.) If you back up videos that are longer than 15 minutes, those files count against your 15GB space quota.

So what happens if you need more space? You can get it — at a price, starting at $4.99 per month for 100GB. You can view the full storage plan and pricing list at https://support.google.com/drive/answer/2375123.

If you don't want to worry about maxing out your online storage space, you can also back up files to your computer (see "Sharing with your computer" earlier in this chapter). Just copy files from your Glass to a folder on your computer.

Accessing backed-up files

You can access synchronized files from the Google+ website on your browser or through the Google+ app on your Android or iOS device. When Glass is connected to the Internet through a Wi-Fi connection, you can also use these services to sync the pictures on your Glass to the Auto Backup album at any time. In addition, you can use the Google+ app to move files from the Auto Backup folder to a Google+ folder that you share with Google+ contacts and/or circles.

Backing up manually (Android and iOS)

After you enable Auto Backup, the Google+ app uploads any pictures or videos you take with your smartphone or tablet to the Auto Backup folder automatically. When you sign out of the Google+ app, any pictures and videos that you take don't upload to the Auto Backup folder. You can upload those pictures and videos later by logging back in to Google+ and tapping Back Up All in the Google+ app screen, shown in Figure 8-13.

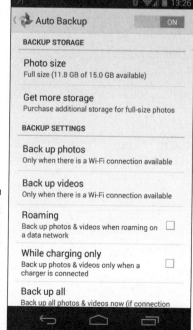

Figure 8-13: The Backup Storage and Backup Settings options on the Settings screen.

Chapter 9

(Almost) Everything Is Possible Online

Answers are what we are trying to get at; search is a process by which you may be able to get answers, but it's not the end goal. It's a mechanism.

— Conrad Wolfram

Searching for information without reaching for your smartphone is one of the big advantages of using Google Glass. It's easy to search the web by using the built-in Google search engine from the Home screen. Glass also offers the Google Now service, an automated personal assistant that shows information cards based on what Glass knows about you and how you use Glass.

In addition, Glass creates timeline entries that are based on how you use Google technologies. If you travel between work and home often, for example, Glass creates a new timeline card with a map automatically; this map shows you the fastest route you can take from your work location back home.

Glass also contains built-in apps for connecting with people. You can use the industry-standard Short Message Service (SMS) to send brief text messages to other users via Glass. You can chat in a live video call in Google Hangouts. And you can send e-mail messages. We show you how in this chapter.

SMS functionality is available only if you're using Glass with an Android smartphone on which the MyGlass app is installed. If you don't have SMS functionality, Glass sends any SMS message by e-mail to the recipient automatically.

Searching to Your Heart's Content

To search the Internet on Glass, you need to be connected to the Internet through the device's Wi-Fi connection or your smartphone's data plan. (See Chapter 6 to find out how to pair Glass with your smartphone.) You have three ways to search for information:

- At the Home screen, say "OK Glass, Google. . ." followed by the word(s) you want to search for or the question you want answered. A couple of seconds after you stop speaking, Glass displays the result onscreen.

- At the Home screen, tap the touchpad and then tap Google on the menu. When the Google screen appears, say the search word(s) or ask your question.

- If the screen is off, you can turn it on by tapping the touchpad three times slowly. When you see the Google screen, say the search word(s) or ask your question.

For every search result, Glass not only displays the result onscreen (in words and/or pictures), but also reads the result to you over its speaker. You don't need to enable any setting to have Glass read the search results to you; this feature works right out of the box. To find out more about Glass basics, including how and when Glass talks to you, see Chapter 4.

What you can search for

You can initiate many types of searches on Glass. Here are a few examples:

- **Service tips:** You can use Glass to calculate the tip when you go out to dinner. If you ask "What is 15 percent of 33 dollars?", for example, Glass displays the result shown in Figure 9-1.

Figure 9-1:
Glass calculates a dinner tip.

✔ **Translations:** You can translate phrases into other languages. When you ask "How do you say 'Good afternoon?' in Slovenian?", you see the translation shown in Figure 9-2.

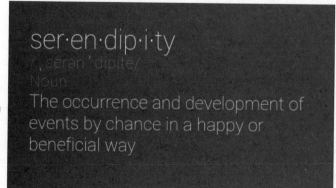

Figure 9-2:
Asking a
question in
Slovenian.

✔ **Definitions:** You may want to get the definition of a word you don't know or want to know more about. When you say "Define *serendipity*," Glass displays the word's definition, as well as its pronunciation and part of speech (see Figure 9-3).

Figure 9-3:
Glass can
be a walking
dictionary.

✔ **Nearby attractions:** You can find attractions in a city you're visiting. To find places to dine in San Francisco, say "Restaurants in San Francisco." Glass finds the restaurants closest to your current location and displays them on a map, as shown in Figure 9-4. You can get more information about a restaurant by selecting the restaurant in the list. (Scrolling through the list may be required.)

Figure 9-4:
Finding food
options
in San
Francisco.

After you search for something on your Glass, the search appears as a card in your timeline (see Chapter 4). You can delete a past search by deleting its card from the timeline.

How to find out more

In some cases, Glass may not give you all the information you want in response to a query, but you don't have to settle for its first answer. You can follow up on a search result in two ways: by asking a follow-up question or by viewing a website.

Ask a follow-up question

You ask a follow-up question by going back to the Google Search screen and asking the question by using pronouns. To get more information about a result that shows you how tall the Eiffel Tower is, for example, you can ask "When was it built?" Glass is smart enough to realize that you're following up on your last question and that *it* refers to the Eiffel Tower, so it displays the screen shown in Figure 9-5, telling you that construction of the Eiffel Tower began in 1887.

View a website

Sometimes, a result screen displays a URL in the bottom-left corner; this URL is a link to the website where Glass got the result information. In Figure 9-6, for example, a search result for a question about why the sky is blue contains a link to a website at the University of California–Riverside.

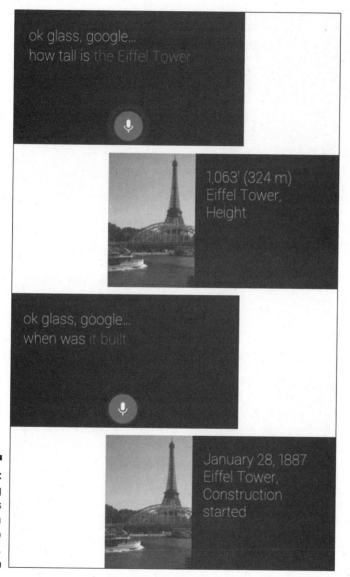

Figure 9-5:
Getting more results from a follow-up question.

To view this website, tap the touchpad and then choose View Website from the menu. The Browser app opens, displaying more information about why the sky is blue. (We talk about the Browser app in more detail in the next section, so read on.)

Figure 9-6:
Some
search
results
contain links
to websites
that pro-
vide more
information.

Browsing Any Time, Anywhere

Google produces its own web browser, Chrome, for desktops, laptops, tablets, and smartphones, so it's no surprise that Google created a special version of Chrome for Glass that takes advantage of the device's user interface.
On Glass, the Chrome browser is called the *Browser app*.

Websites designed for mobile devices look best on Glass, which is also a
mobile device and has screen resolution (640 by 360 pixels) similar to that of
many other mobile devices. The Browser app makes it easy for you to navigate all types of websites that you encounter, however.

If a website has a mobile version, that website detects that Glass is a mobile
device and opens the mobile version automatically. The Browser app on your
Glass, however, doesn't work with gestures that you can use on mobile websites. On a smartphone, for example, you can view different sections of a site
by swiping left to right. Because you need to use gestures to navigate in the
Glass operating system, however, you need to tap the section heading link to
view more information.

You can open the Browser app in a search result screen by tapping the touchpad and then choosing View Website from the menu. The Browser app displays
the website and highlights the center of the screen (see Figure 9-7), which
shows the area of the site that you can manipulate to do various things.

You can use your finger and your head to move around a website by using
four gestures: scrolling, zooming, panning, and tapping. We discuss these gestures in the following sections.

Figure 9-7:
The center of the screen contains a highlighted area bounded by a dashed circle.

The Browser app doesn't close automatically after a period of inactivity. Like any app, the Browser app drinks juice from your battery, and if you decide to do something else on Glass, the Browser app remains active. So to prevent any unexpected low-battery message from popping up on the screen (which, of course, will happen when you're doing something important), be sure to close the Browser card in your timeline. If you need to know how to do that, get up to speed in Chapter 4.

Scroll up and down

To scroll down a web page, slide your finger forward on the touchpad, and to scroll up a web page, slide your finger backward on the touchpad (see Figure 9-8). As you move up and down the website, the viewable area onscreen changes as well.

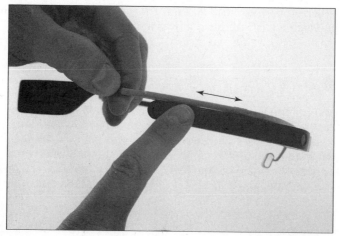

Figure 9-8:
Scrolling a web page.

@Wiley/Michael E. Trent

Zoom in and out

Slide two fingers forward or backward to zoom in and out, respectively, as shown in Figure 9-9. When you zoom in, the text and images onscreen and in the viewable area get larger. When you zoom out, the text and images onscreen and in the viewable area get smaller.

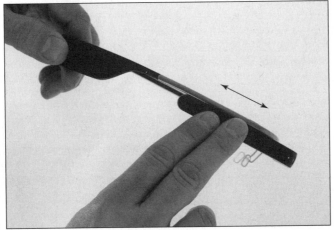

Figure 9-9: Zooming a web page.

@Wiley/Michael E. Trent

Look around a web page

If a web page is wider than the Glass screen (which may be the case more often than not), just use your head — that is, hold two fingers on the touchpad and then move your head around. The web page moves up, down, left, and right onscreen as your head moves in those directions.

Tap a link

In the center of the viewable area is a group of small dots arranged in a circular pattern (see Figure 9-10). This *circular pointer* serves the same function as a mouse pointer on a desktop computer. To position this pointer, maneuver the screen so that the pointer appears over the hyperlinked text or image that you want. When you do, the text link changes color or a colored box appears around the image. To select the link, tap the touchpad.

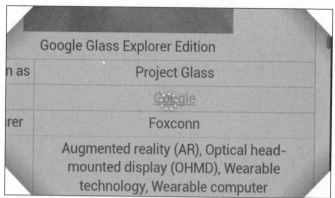

	Google Glass Explorer Edition
n as	Project Glass
	Google
rer	Foxconn
	Augmented reality (AR), Optical head-mounted display (OHMD), Wearable technology, Wearable computer

Hanging Out with Your Friends

Because Glass comes from Google, several Google technologies work with it right out of the box. One of these technologies is Hangouts, Google's video-conferencing app. Glass lets you use Hangouts to send and receive chats and photos, as well as make video calls.

If you want to have a video call through Hangouts, your contact must also be logged in to Hangouts and connected with you as a friend.

You can initiate a video call with a contact on Google+ or a group of contacts in a Google+ circle. You can also receive a video call from another Google+ user (or even from another Glass user). When you participate in a Google Hangout, you see the video of the other participants on your screen, and the other participants see your point of view — that is, what you're seeing through the camera of your Glass.

Keep in mind four important facts about receiving and initiating video calls on Glass:

✔ You can make video calls only if you're running the Hangouts app on your Glass.

✔ Video calls use up your battery a little faster than other activities do, so it's best to keep your video conversations as brief as possible.

✔ As more invitees join your Hangout, the faces on the screen become smaller to accommodate all the people you're chatting with. Don't worry, though; the number of faces you see won't resemble a CNBC television discussion with multiple analysts, in which you can barely see all the faces on the screen. You'll still be able to recognize everyone in the Hangout.

✔ We recognize that you're an upstanding citizen, but this reminder is important: Consider etiquette and common sense when you're in a video Hangout. Refrain from broadcasting events or content where such activity is prohibited, such as a movie playing in a theater.

Joining a Hangout

When you get a request to join a Google Hangout, you hear a chime on the Glass speaker and also see a picture of the person who's inviting you to the Hangout.

If you want to join the call, tap the touchpad or say "OK Glass, accept call." To decline the invitation, ignore it by swiping forward on the touchpad or say "OK Glass, reject call."

Starting a Hangout

You can start a Hangout of your own from any screen. Just say "OK Glass, make a video call to. . ." followed by the name of the contact or the name of the Google+ circle containing the contacts you want to invite to the Hangout. If the contact has logged in to the Google+ website on his or her computer or has the Hangout app for the iOS or Android mobile platforms, that contact can participate in the Hangout.

If the contact or circle you name isn't in your Google Contacts database, a brief message appears onscreen, stating that the contact or circle doesn't exist. In such a case, you have to add that contact first (see Chapter 5).

After you invite a contact to a Hangout, you see the Waiting screen, which tells you that you're waiting for your invited contact to join your Hangout (see Figure 9-11).

You can continue to do other things while you wait for a contact to join your Hangout (see the next section) by swiping down on the touchpad to access your timeline.

When someone joins the Hangout, you hear the chime that sounds when you receive a request to connect; then you see the contact's face on the Hangouts screen. When more of your invitees connect with you in the Hangout, you hear that chime and see those contacts' faces onscreen.

Figure 9-11:
The Waiting
screen
keeps you
company
while you
wait for
contacts to
arrive.

Taking action in a Hangout

While you're in a Hangout, you can perform other tasks, such as check your timeline for an appointment on your calendar. When you do, your camera is muted (that is, turned off), and other Hangout participants won't be able to see your point of view. Your audio is still on, however, so your Hangout participants will be able to hear you. You can re-enter the video call and unmute your camera by tapping the active video-call card on the left side of the Home screen.

During your Hangout, you can do a variety of other things by tapping the touchpad and then choosing one of these options from the menu:

- **People:** See a list of people who are in the video call with you.

- **Leave:** Exit the Hangout. If you're the person who started the Hangout, the Hangout ends for all participants when you leave.

- **Invite:** Add people or circles up to a maximum of ten contacts.

- **Mute Mic:** Deactivate the microphone so that Hangout participants can't hear you.

- **Mute Video:** Deactivate the camera so that Hangout participants can't see what's on your screen. (You may want to participants know that you're turning off your video before you do this so that they don't wonder what happened to you.)

Sending E-mail Messages

E-mail is a very popular way of sending information, so Glass lets you send and receive e-mail messages. If your contact has only an e-mail address and no phone number or any other contact information, such as a Google account, Glass sends messages for that person to his or her e-mail address automatically.

Before you send e-mail messages on your Glass, you need to add e-mail contacts via the MyGlass website or the MyGlass app on your smartphone. For details, see Chapter 5.

You've got mail

When you receive an incoming e-mail message, Glass plays a unique chime, even if the display is off. Here's how to read and respond to e-mail:

1. **Turn on the display by tapping the touchpad or using Head Wake Up (see Chapter 6).**

   The e-mail card appears, as shown in Figure 9-12, including the e-mail Subject line, a snippet of the message, and photos of you and/or the recipient (if those photos exist).

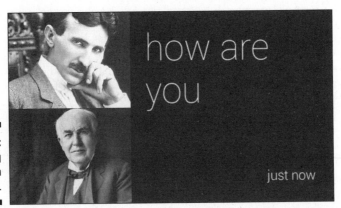

Figure 9-12: Getting e-mail on Glass.

2. **Tap the touchpad and then choose one of the following options from the resulting menu:**

   - *Read More:* Read more of the message if it's too long to fit on the screen.

   - *Reply:* Open the Reply screen so that you can speak a reply and then send it to the recipient.

   - *Archive:* Save the message to Glass and delete the message from the timeline.

   - *Read Aloud:* Have Glass read the displayed message snippet to you.

   - *Star:* Mark the message as one of your favorites.

   - *Delete:* Delete the message from the timeline.

3. **Swipe down on the touchpad to close the menu.**

4. **To close the e-mail card and return to the timeline, swipe down on the touchpad while you're viewing the e-mail card.**

If your e-mail message is part of a *thread* (related messages about the same topic), the e-mail message card appears as a bundled card. You can read all the messages in the thread by tapping the card and then swiping backward and forward on the touchpad. We discuss bundled cards in Chapter 4.

Open hailing frequencies

It's easy to send an e-mail message to someone else. Just follow these steps:

1. **Say "OK Glass, send a message to."**

   Glass displays your contacts list.

2. **Speak the name of the intended recipient or select it by using the touchpad.**

   The Send Message To screen appears, as shown in Figure 9-13.

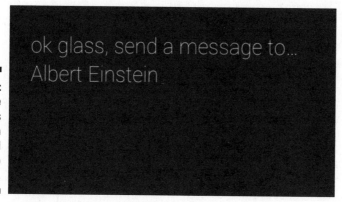

ok glass, send a message to...
Albert Einstein

Figure 9-13: Confirm the recipient's name in the Send Message To screen.

3. **Speak your message.**

   As you do, the text appears below the recipient's name, as shown in Figure 9-14.

A couple of seconds after you stop speaking, the Glass screen shows a message for a few seconds so that you can cancel the send by swiping down on the touchpad. If you do nothing, you see the `Sent message` text on the screen for a second or two; then the message appears as a card on your timeline. The recipient sees the words `Sent through Glass` appended to the end of the message.

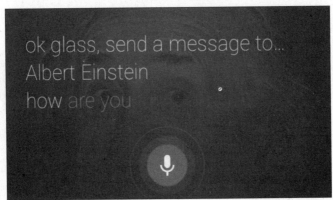

Figure 9-14:
The name of your recipient appears onscreen as you dictate your message.

Getting Fast Facts with Google Now

You may have heard about automated personal assistants such as Siri, the iOS feature that lets you ask your device questions and get spoken information about all sorts of stuff. Google has its own automated personal assistant, called Google Now. The Google Now app gives you the information you need in several categories, including Calendar, Flights, Weather, Sports, Stocks, and Places (see Figure 9-15).

The Google Now app monitors how you use Glass and places cards on your timeline that reflect your interests. If you're traveling, for example, Google Now detects flight delays and displays a card onscreen automatically. In some cases, you may find out about a delay before you leave for the airport.

To set up Google Now on Glass, you need an Android smartphone, because the Google Now app requires the Google Search app that comes preinstalled on Android smartphones running the Android 4.1 (Jelly Bean) operating system or a later version.

Figure 9-15:
Google Now
cards.

Setting up Google Now in Google Search

First, set up Google Now in the Google Search app by following these steps:

1. **Open the Google Search app on your Android smartphone.**

2. **Tap Search on the menu.**

3. **In the Google Now section of the Search screen, tap the button to change it from Off to On.**

4. **In the confirmation screen, tap the Yes, I'm In button.**

Setting up Google Now on your Glass

With that out of the way, here's how to set up the Google Now app on Glass:

1. **Open the MyGlass app on your smartphone.**

2. **Tap Glassware on the menu at the top of the screen.**

3. **Locate and select the Google Now card.**

4. **Tap the Off button in the top-right corner of the Google Now screen.**

   The Off button changes to On. On the Glass screen, you see the Google Now card on the timeline, ready for use.

Chapter 10

Using Glass with Your Phone

To me, e-mails are a little bit frustrating. I think that the telephone is much preferred because you get the sound of the voice and the interest and everything else you can't see in an e-mail.

— T. Boone Pickens

Glass doesn't have a phone incorporated into its hardware — at least, not in the first version of the device — so it relies on any smartphone that allows you to make Bluetooth connections. When you connect your smartphone and your Glass, you'll be able to place calls by speaking into Glass and also receive calls over your hands-free handset.

If your contact has a phone number, and you want to send a message to that person, Glass sends a Short Message Service (SMS) message to the person's smartphone or tablet. The SMS service works with your Android smartphone so you can tell Glass to dictate a message to a specific contact. When you send that message, it appears in the contact's messaging app on his smartphone.

Finally, the MyGlass app available on smartphones comes with a nifty feature called *screencasting,* which allows you to project what's happening on the Glass screen to a projector or a smartphone. You can even use the MyGlass app to control Glass from your smartphone while you're screencasting, which is useful for explaining how to navigate on Glass to your audience.

Making and Taking Phone Calls

The benefit of using Glass as a headset for making and receiving calls is that you don't have to fumble for your phone in your pocket and then keep looking down at your phone or holding the phone to your head to converse with someone. Instead, Glass gives you the freedom to make and receive calls hands-free.

For purposes of this chapter, we assume that you've already connected your smartphone to your Glass via a Bluetooth connection. If you aren't sure how to do that, see Chapter 6.

Opening a channel

It's easy to make a call. Just say "OK Glass, make a call to. . ." followed by the name of the person in the Glass contact database whom you want to call. If you say "OK Glass, make a call to Albert Einstein," you see the call screen shown in Figure 10-1.

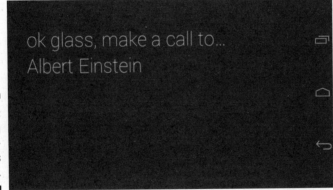

ok glass, make a call to...
Albert Einstein

Figure 10-1:
Calling
a world-
famous
scientist.

Be sure that you have a phone number in the contact record on your Glass for the person you want to call (see Chapter 5).

Captain, you're being hailed

When you receive a phone call, Glass plays a rising and falling chime on its speaker and then displays the name and phone number of the person who's calling you.

To accept the call, do either of the following:

- ✔ Tap the touchpad and choose Accept from the menu.
- ✔ Say "OK Glass, answer call."

To reject the call, do one of the following:

- ✔ Tap the touchpad and swipe down.
- ✔ Tap the touchpad, swipe forward, and then choose Reject from the menu.
- ✔ Say "OK Glass, decline call."

Controlling your call

During a call, you can use either your Glass or your smartphone to control the call, as follows:

- ✔ On your Glass, to mute your voice so that the other person can't hear you, tap the touchpad and choose Mute Your Mic from the menu. You can unmute your voice and be heard again by tapping the touchpad and choosing Unmute from the menu.
- ✔ On your Glass, end the call by tapping the touchpad and then choosing End Call from the menu.
- ✔ On your smartphone, end the call by tapping End onscreen.

Sending SMS Messages

If a contact record stored in the contact database on your Glass includes a phone number, you can send an SMS message to that contact by using your voice instead of hunting and pecking with your thumbs as you would on a smartphone.

You can access your imported contacts from the Google Contacts database only by using the touchpad; you can't use the voice feature. That is, when you send a message, you need to scroll through the contact list by swiping backward or forward on the touchpad. When you find the contact name you want, tap the touchpad.

Glass doesn't have built-in SMS messaging capabilities — at least, not yet. Instead, it uses the Google Voice app on your phone (if you've installed it and set it up) or the messaging app on your smartphone to send SMS messages.

At this writing, SMS messaging is available only via a Bluetooth connection to an Android smartphone running the MyGlass app (see Chapter 5).

Sending to a Glass contact

To send an SMS message, follow these steps:

1. **Say "OK Glass, send a message to. . ." followed by the name of the contact.**

   The Send Message screen appears, as shown in Figure 10-2, so you can see that Glass has recognized the name of your contact.

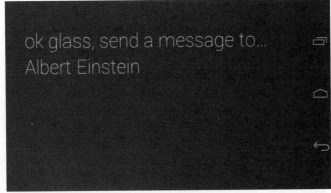

Figure 10-2: The name of the recipient appears on the Send Message screen.

2. **Start dictating your message.**

   As you speak, the message appears below the contact's name (see Figure 10-3).

   A couple of seconds after you stop speaking, Glass sends the message. You see Sent text onscreen for a second or two; then the message appears in a card on your timeline.

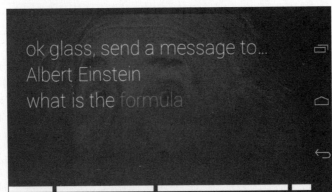

Figure 10-3: Dictating an SMS message.

You have only a couple of seconds to cancel a message before it's sent off to the contact's message app, so keep your messages short and to the point.

Sending to a smartphone contact

If you want to send an SMS message to someone who isn't in your Glass contacts list but is in your smartphone's contacts list, here's how to do that:

1. **On the Home screen, tap the touchpad.**

2. **Choose Send Message from the menu.**

3. **Find the smartphone contact by swiping backward and forward on the touchpad.**

4. **Dictate the message (see "Sending to a Glass contact" earlier in this chapter).**

Going Live with Screencasting

Your Glass can use your smartphone to show people a live view of what you're seeing on the Glass screen. You can show people what you're looking at on your smartphone, for example, or you can connect Glass to a large monitor or even a projector by connecting the smartphone to the monitor or projector (via an HDMI adapter cable). Before you can screencast, you need to set up the MyGlass app on your smartphone (see Chapter 5).

Here are some important screencasting facts to keep in mind:

✔ Screencasting doesn't transfer sound to your phone; the sounds you hear on your Glass stay confined to the Glass speaker. If you're presenting Glass to an audience, make sure that you tell them what you're hearing on your Glass as you perform tasks.

✔ When you begin your presentation, tell your audience that there'll be a brief delay before what you see on the Glass screen appears on the smartphone, monitor, or projector. This disclaimer helps your audience members understand that the slow performance they see on the phone, monitor, or projector doesn't match what they'd see on the Glass screen.

Starting and stopping a screencast

When you've set up MyGlass as described in Chapter 5, you can screencast by following these steps:

1. **Open the MyGlass app on your smartphone.**

2. **Choose Screencast from the menu (see Figure 10-4).**

   The screencast begins immediately, and the phone's screen displays exactly what you're seeing on the Glass screen.

3. **To end the screencast, tap the smartphone's Back or Home button.**

Figure 10-4: Choose the Screencast option to start screen-casting.

Connecting to a video device

HDMI cables allow you to connect your Android smartphone to any video device that has an HDMI port, including most monitors and projectors produced in the past two or three years.

Because each phone is different, be sure to consult your smartphone's documentation and the manufacturer's website to find out how to connect your smartphone to your monitor or projector effectively.

Controlling a screencast

If you're using a smartphone running Android 4.1 or later, you can use the MyGlass smartphone app to control your Glass. (Remote control of your Glass isn't available in the iOS MyGlass app as of this writing.)

You may want to control your Glass remotely for several reasons:

- ✔ You feel more comfortable using your smartphone to do things on your Glass.

- ✔ You want to transition to the Glass user interface slowly.

- ✔ You want to use your smartphone to show someone else how to navigate the Glass screens while you're using Glass.

After you begin a screencast, the smartphone screen displays what's on the Glass screen, as shown in Figure 10-5.

Figure 10-5:
The Glass screen as it appears on a smartphone.

TIP

Best and worst bets for screencasts

If you're wondering how to make Glass shine for your audience, here are some Glass features and tasks that make good screencasts, as well as some that don't (at this writing, anyway).

Good:

✔ Swipe back and forth on the timeline, open a card, and open a card bundle.

✔ Open the Google Now card and point out Google Now features, such as the latest appointments on your calendar.

✔ Make a phone call to someone in your audience. (Be sure to add that person to your contacts list before your presentation. Also, test a phone call first to ensure that the call video appears properly in the screencast.)

✔ Send an e-mail or SMS message to someone in your audience. The audience will see the message as you speak into Glass, and the person who receives the message can verify that it arrived on his or her smartphone.

✔ Show the current weather where you are, or ask an audience member for a city and then show the current weather and forecast for that city.

✔ Take a picture of something — even of your audience members, if they agree to let you — and share that picture on your favorite social network.

Not so good:

✔ Web browsing can be a bit sluggish, so test it at the site of your presentation beforehand. If the test fails, you can easily keep browsing out of your presentation.

✔ If you turn or move too quickly while demonstrating how GPS works, the screencast starts to lag. Test this feature before the presentation to get an idea of how quickly you can move without any problems.

✔ Because of the delay between Glass and the projector, you may want to reserve video demonstrations for people who also have Glass on their heads.

You can use the following gestures on your phone as equivalent gestures on your Glass:

✔ Swipe up and down on the smartphone screen to swipe up and down, respectively, on the Glass touchpad.

✔ Swipe left and right to swipe backward and forward, respectively, on the Glass touchpad.

✔ Tap the smartphone screen to tap the Glass touchpad.

If you decide that you don't want to use your smartphone to control Glass any longer, just tap your smartphone's Back or Home button.

Chapter 11

Other Cool Stuff

This wired generation is kind of cool.

— LeVar Burton

One of the best things about owning any new device, especially Glass, is discovering all the cool stuff it can do. In this chapter, we show you a few fun things you can do with your Glass.

Navigating the World in a Flash

Before you try to get directions on your Glass, be sure that you've done three things:

✔ Installed the MyGlass app and enabled GPS location services on your smartphone (see Chapter 5)

✔ Paired Glass with your smartphone (see Chapter 3)

✔ Connected Glass to the Internet (see Chapter 3)

Getting routed

When everything's in place, give the built-in navigation app a try by asking Glass for directions to a major city. By default, the directions you see are those for driving to your destination.

The Google Maps view is good, but it doesn't give you precise directions. To augment a map with a route, for example, say "OK Glass, get directions to San Francisco." The route information card appears, as shown on the right side of Figure 11-1. This card lists your current location, the name of the next street or exit you need to take, the turn direction, and the distance to the turn, as well as how long it will take to get to your destination.

Figure 11-1: The route information card does almost everything but drive the car for you.

To close the Google Maps or route information screen (see the next section), just tap the touchpad and then choose Stop from the menu that appears.

Even if you take your Glass off your head or turn the screen off while the navigation app is running, the app continues to run in the background, and whenever new directions occur, Glass plays those directions over its speaker.

Deciding how to get there

You don't necessarily have to drive to use Glass for navigation. The device can give you directions no matter how you're traveling, be it on foot, via bicycle, on public transportation, or by car.

Google built the Google Maps Transit system into Glass so that users in select cities can get information on local bus and train travel. You can view train and bus timetables, find out whether (and when) to switch trains or buses, find the distance to the nearest terminal, and see how long it will take you to reach your destination based on your method of travel. You can see which cities are served by Google Maps Transit at www.google.com/landing/transit/cities/index.html.

Change your travel mode

If you want to change your method of travel, you can do so from within the Google Maps screen or the route information screen. Just tap the touchpad and then choose Transit (for bus or train), Drive (for car), Walk, or Bike. The information onscreen changes to reflect the new mode of travel.

The next time you view directions in the Google Maps or route information screen, the mode of travel defaults to your last selected mode.

View your entire route

If you want to see the route for your entire trip, here's what to do:

1. **Tap the touchpad.**

2. **Select the Show Route option by tapping the touchpad.**

   The Show Route screen displays a map that shows your starting point (a blue arrow), the path to your destination (a blue line), and the destination (a red pin), as shown in Figure 11-2. Below the map, you also see the following information, from left to right:

   • An icon representing the selected mode of transportation

   • The destination city

   • The miles from your current location to your destination

   • The travel time from your current location to your destination

Figure 11-2:
Check your mileage and travel time (in red).

Review transit directions

If you're traveling to your destination via one or more transit modes, you can get step-by-step directions to your destination. You see details about your trip on a screen that's similar to Figure 11-3.

Figure 11-3:
The directions
include how
long your
entire round
trip will
take.

Here's what these symbols represent:

- ✔ **Travel methods:** The white icons above the colored line show the methods of transportation you'll need to take. You may see icons that denote walking (a stick figure), taking a bus (a bus icon), taking a train (a train icon), biking (a bike icon), and/or driving (a car icon) during different phases of the trip. The trip depicted in Figure 11-3, for example, has seven phases. You'd walk to the bus station, take several bus routes, and then take the city light-rail system to your destination.

- ✔ **Trip phases:** The line denotes what you'll be doing at any phase of the trip and is color-coded to correspond with the icons above the line. Here's what the various segments of the line shown in Figure 11-3 mean:

  - *Cyan dotted line:* This line represents a walking route.

  - *Blue solid line:* This line shows a bus or light-rail route.

    You may see different colors for different types of routes. In Figure 11-3, the light blue line signifies the bus route in your current city, and the dark blue line signifies the bus and light-rail routes to and within your destination city.

  - *Green solid line:* This line depicts a train route (not shown in Figure 11-3).

- *Circles:* A circle marks the beginning or end of a phase. The colors of the circles denote what you're going to do next. A blue circle, for example, means that the next step in your trip is boarding a light-rail train.

- *Lines connecting circles:* These symbols represent the length of each phase of the trip relative to the overall length of the trip.

Checking out the neighborhood

Because Glass is connected to the GPS locator on your smartphone, your phone keeps track not only of where you are, but also of what interesting attractions are nearby. After all, Google doesn't want you to miss any photos you can take with Glass and share with your family members and friends.

When an attraction is near you, a card on the timeline shows that attraction. If your phone finds several attractions in your area, you see a card bundle, as shown in Figure 11-4.

Figure 11-4: When you're in Paris, attractions are all around you.

View the attractions as individual cards by tapping the bundle or (if only one attraction is nearby) view the attraction's card in the timeline (see Figure 11-5).

Figure 11-5: Get details on a notable photo spot in Mountain View, California.

Directing your attention to appointments

You may find that you're so enamored with getting directions and finding out more about your environment that you forget about your next meeting. Thanks to the connection between your smartphone and your Glass, you can get reminders of upcoming appointments that you've stored on the Google Now service. For information on how to set up reminders on Google Now, see `https://support.google.com/websearch/answer/3122344?hl=en`. You can also find out more about setting up and using Google Now on Glass in Chapter 9.

If you want to find certain types of attractions or businesses, such as a coffee shop where you can relax and catch up on your e-mail, just say something like "OK Glass, get directions to a coffee shop." The resulting Google Maps page shows a list of all the coffee shops near your current location. You can scroll up and down the list of coffee shops and then tap a name in the list to get directions to that coffee shop.

Finding Out What's Playing

Another good use of Glass is listening to music through the device's speaker — much easier than digging up your smartphone and headphones and then putting your smartphone back where you found it. But you can do more than just listen to music. The Sound Search feature makes it easy for you to ask who's singing that song.

While you're listening to a song that's playing over the speakers at your favorite coffee shop, you can find out the name of that song and the artist in one of two ways:

- Slowly tap the touchpad twice until the Google Search screen appears; then swipe forward on the touchpad.
- Say "OK Glass, Google what song is this?"

Whichever method you use, Glass listens to the song and then displays the details (see Figure 11-6). Now you can rest easy at night.

Figure 11-6: Glass identifies the song name and artist(s).

Treating Your Eyes Right

When you're outdoors, you may find yourself squinting because the light is too bright. Not to worry. Google includes a sunglass shade with every Glass. You can also use a different frame if you don't like the one that came with your device. We show you how to customize your look in this section.

You can get two types of eye protection for your Glass. The first type is a clear shield that protects your eyes, akin to safety goggles (but much more stylish and comfortable). You can also use a sunglass shade, which helps block sunlight and protects your eyes on a bright, sunny day. Both clear and sunglass shields come with Glass.

Throwing shade

Google partnered with Maui Jim (www.mauijim.com) and Zeal Optics (www.zealoptics.com) to put both a sunglass shade and a clear shield inside the box your Glass came in. Here's how to attach either one:

1. **Align the shade outward so that it wraps around the front of your frame, as shown in Figure 11-7.**

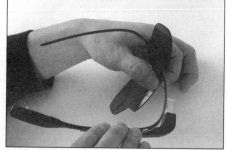

Figure 11-7: Align the shade.

@Wiley/Michael E. Trent

2. **Tilt the shade so that the black cushion along the top of the shade's nose bridge slides over the nose stems (see Figure 11-8).**

Figure 11-8:
Tilt the shade.

@Wiley/Michael E. Trent

3. **Slide the shade into place (shade outside the nose stems and black cushion inside the nose stems), as shown in Figure 11-9.**

When the shield is firmly positioned and flush with the top of the Glass frame (see Figure 11-10), you're ready to go outside. Enjoy the sunshine.

Figure 11-9:
Position the shade along the top of the frame.

@Wiley/Michael E. Trent

Figure 11-10:
Shield in position.

@Wiley/Michael E. Trent

If you want a clear shield to protect your eyes from the elements, you can find one (and other Glass accessories) at `https://glass.google.com/getglass`. If you don't see the desired accessory, check back in a few days.

Switching frames

The frame that came with your Glass may not be exactly to your liking. If not, you can purchase a new frame for $225 (plus tax) at `https://glass.google.com/getglass`. With your new frame, you also get a hard-body case, screwdriver, cleaning cloth, and extra nose pads.

What's more, Google has now made four titanium frames available with your prescription lenses so you can see clearly and get the advantages of Glass. You can see all the frames at `www.google.com/glass/start/how-it-looks`.

If you want to affix prescription lenses to your titanium frame, you need to talk with your eye care provider and/or visit a Glass Preferred Provider to have him or her install your lenses. Google recommends lenses cut within +4 to –4 with astigmatism up to 2D. You can find a list of Glass Preferred Providers at `www.google.com/glass/help/frames/providers`.

You should check with your vision insurance company to see whether the company will cover the cost of your new Glass frame with prescription lenses. At this writing, only Vision Service Plan has announced support for the new Glass titanium frame.

Off with the old

As soon as you get your new frame, remove the old one by following these steps:

1. **Remove the screw on the inside of the device with a Torx T4 screwdriver.**

   You can purchase a Torx T4 screwdriver at any hardware store (online or bricks-and-mortar).

2. **Grip the frame with your left hand and the pod with your right hand.**

   Google refers to the side of Glass that contains the camera, screen, speaker, and other circuitry as the *pod*.

3. **Gently lift the pod off the frame.**

   You see the Federal Communications Commission (FCC) identification number on the inside of the frame.

On with the new

To attach the new frame, follow these steps:

1. **Grip the frame with your left hand and the pod with your right hand.**

2. **Align the screw hole in the frame with the screw hole on the pod.**

3. **While holding the frame and pod together with your right hand, place the screw in the frame hole, and use a Torx T4 screwdriver to screw the frame and pod together.**

Part IV
Give More Power to Your Glass

In this part . . .

✔ Find out how to get third-party apps for Glass.

✔ Understand how to check and upgrade the Glass operating system.

✔ See how to hack Glass so you can get the most out of its apps and performance.

✔ Discover how to add existing Android apps to your Glass at www.dummies.com/extras/googleglass.

Chapter 12

Finding Third-Party Apps

· ·

· ·

Technology, and applications of this technology, will continue to improve and evolve, providing unprecedented, global access to information, individuals, training, and opportunities.

— Maynard Webb

*T*he promise of wearable computers is that the devices themselves go away. That is, they melt into the background to deliver data as you need it instead of requiring you to look for something, as other computing devices do.

Apps written specifically for Google Glass are available now. These apps, which Google calls *Glassware,* go a long way toward fulfilling the promise that you see data only when you need it.

Shop for Glassware

Apps aren't installed on Glass by default. Instead, you shop for them on the MyGlass website or on the MyGlass app on your smartphone (see Chapter 5).

Browse apps

To check out some of the Glassware that's available, follow these steps:

1. **Go to the MyGlass website (www.google.com/myglass) on your smartphone or computer.**

2. **Click or tap Glassware on the menu bar at the top of the MyGlass home page (see Figure 12-1).**

Figure 12-1:
Select
Glassware.

The Glassware page appears (see Figure 12-2), displaying two sections of apps that are available for Glass. The top section, Featured Glassware, lists apps that Google recommends. Below it, the All Glassware section shows all available apps. In both sections, apps are represented as cards and listed alphabetically from left to right.

Each app card contains the following information:

• The name of the app

• A description of the app

• The app icon

• A blue icon with a check mark (if the app is installed on your Glass) or a gray icon with a plus sign (if the app isn't installed)

3. **Scroll the screen to view all the apps.**

Many apps are available at this writing, including the following:

✔ **Allthecooks Recipes:** If you not only like to cook, but also like to share recipes, tips, and ask questions of other chefs, this app gives you the best of both worlds. You can read and follow recipes on your Glass instead of flipping through a printed recipe book and also connect with fellow app users to get recipes, share information, and ask how you can make your dishes even better.

✔ **Compass:** This app tells you the direction in which your head is pointing so you can orient yourself in the real world. The Compass app also uses your GPS location when you have your Glass connected to your smartphone to tell you when you're near an interesting landmark, in case you want to check it out.

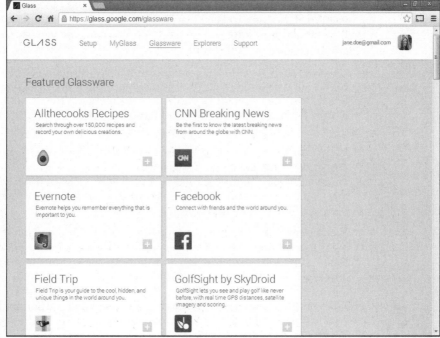

Figure 12-2:
The Glassware page displays available apps for Glass.

✔ **GolfSight:** Glass is great for golfers, because you can use the device without letting go of your club to check your smartphone. What's more, the GolfSight app for Glass lets you calculate the distance to the next hole and adjust your strategy accordingly.

✔ **IFTTT:** Use the IFTTT app (the acronym stands for *If This, Then That*) to tell your Glass to do something when an event is triggered. You can tell IFTTT to send you an e-mail message whenever you're mentioned in a Twitter post, for example.

✔ **Mini Games:** Google created five simple games — Balance, Clay Shooter, Matcher, Shape Splitter, and Tennis — that let you have fun and take advantage of the unique features of Glass. All you have to do to start playing these games is say "OK, Glass, play a game" at the Home screen.

✔ **Spellista:** If you're a big fan of word-jumble games, the Spellista app is for you. Each level of the game allows you to unscramble different words. You can also play the game with other Spellista users through your Internet connection.

✔ **Stopwatch:** When you need to time a process or an event, you can use the Stopwatch app for Glass instead of using your smartphone or even your watch.

✔ **Strava:** Going out for a run or a bike ride? Put Glass on your head and use the Strava app to keep track of statistics you want to know, such as how far and how long you've traveled. You can also view a map showing where you are on your route.

✔ **Umano:** Every day, this app sends your Glass interesting articles on topics such as technology, politics, and getting more from your personal life. The articles appear as cards on the timeline. Professional voice actors — not computer voice simulations — read the articles to you on your Glass, so the app sounds like a personal radio broadcast.

✔ **Video Voyager:** The Video Voyager (VV) app allows you to associate videos you take on your Glass with the locations where they were shared. You can share your videos taken with VV with other VV users, and you receive notifications when other VV users share videos taken near your current location.

✔ **Word Lens:** When you're in a foreign country, viewing printed text or signs in a different language, you can use the Word Lens app to view those words and have the app translate them into your preferred language.

If you're not sure which apps are worth using, and you'd rather not take the time to download and test each one, you can see what apps other Glass users are using on the Google+ website. See "Get Some Friendly Advice" later in this chapter.

Download apps

As of this writing, all Glass apps are available for free. When you find an app that you're interested in, you can download it by following these steps:

1. **Click or tap the app's card (refer to Figure 12-2).**

   A plus icon appears in the bottom-right corner of the card (see Figure 12-3).

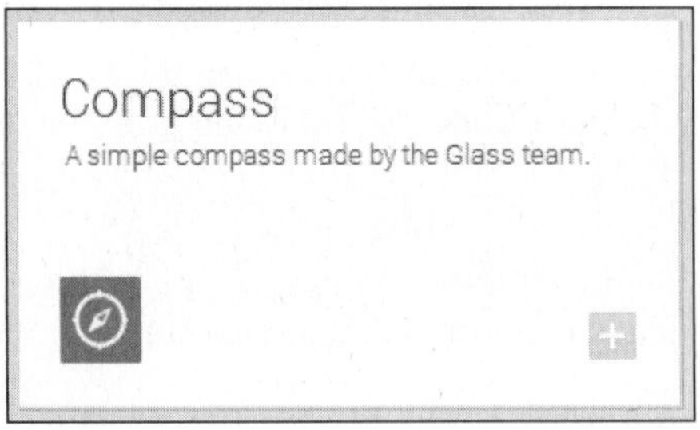

Figure 12-3:
Click or tap
the plus
icon on the
card.

Compass

A simple compass made by the Glass team.

2. **Install the app by clicking or tapping the plus icon.**

   The app's description card appears (see Figure 12-4).

3. **To see what permissions you need on your Glass to run the app, click or tap the Permissions link.**

   A permissions tab or window pops up. To close the tab or window, click or tap the Close button (X) in its top-right corner.

4. **Do one of the following:**

   • To close the card without installing the app, click or tap its Close button.

   • To install the app, click or tap the Off button.

During installation, you see the progress of the installation on the card; then the new app's card appears on your timeline, as shown in Figure 12-5. When you see this card, you know that installation was successful and that you can start exploring the new app on your Glass.

Start Me Up

When you have a new app on your Glass, you can start it from the timeline in either of two ways:

- ✔ If your new app accepts voice commands, say "OK Glass. . ." You see a menu of commands for starting various apps. You can scroll through the list by moving your head up and down until you see the command you want. Start the app by speaking the command you see onscreen.

- ✔ Tap the app's card in the timeline (refer to Figure 12-5).

If you download and install the Compass app, for example, you start it from the timeline by saying "OK Glass, start Compass." The Compass app screen appears, showing the direction in which you're facing (see Figure 12-6).

Figure 12-6: The Compass app shows that you're facing north.

Get Some Friendly Advice

At this writing, the Glassware page of the MyGlass website doesn't include user reviews. But there's an easy way to find out what apps other Glass users are using and what they think of those apps: Visit the Google+ website at http://plus.google.com.

When a Glass user shares pictures or video of a running Glass app in a Google+ post, the #throughglass hashtag is added to the post automatically. So the next time you visit Google+ on your computer or smartphone, search for #throughglass, and you'll find a lot of posts by other Glass users (see Figure 12-7).

Figure 12-7:
A post of
a picture
taken
on Glass
contains the
#through-
glass
hashtag.

If you want to connect with a specific user after reading his post about a Glass app, you can follow that user by clicking or tapping his name in the post and then clicking or tapping Add to Circles on his profile page.

Chapter 13

Upgrading Your Glass

> *Just as established products and brands need updating to stay alive and vibrant, you periodically need to refresh or reinvent yourself.*
>
> — Mireille Guiliano

*J*ust like any other piece of technology, Glass isn't a piece of artwork that's meant only to be stared at through a protective translucent box. (If you want to do that, of course, feel free.) Glass is designed to be used — and used so that it enhances the life of the person who wears it.

The old promise that technology will make our lives easier doesn't mean that we don't have to take time out of our lives to get the latest operating-system (OS) updates that add new features and fix bugs. At least Google has made updating as easy and unobtrusive as possible. You can do the job yourself or let Glass do the work, as you see in this chapter.

Whether you update manually or automatically, never power off Glass during an update process. If you do, when you turn your Glass back on, you may find that the OS isn't working properly, and you may also have lost data stored on the device.

Letting Glass Update Itself

Like any other modern OS, the Glass OS can download updates by itself. Glass connects to the Google servers periodically for updates, and if it finds one, the device downloads the update automatically.

Your Glass won't install updates automatically, however, unless Wi-Fi is enabled and the device is charged. When you plug Glass into the charger, the device updates the OS (if an update is available) while it charges the battery.

Before you charge the battery, be sure that the display is off and that you have a Wi-Fi connection enabled (such as through a wireless router on your home network). If Glass is off, the device will turn on when you start to charge it.

Glass installs OS updates as soon as its battery is 50 percent charged, so you don't need to worry about fully charging Glass to get updates.

Finding out what got installed

New OS updates may include some new features that you see right away. Other improvements happen under the hood and aren't readily apparent. Google gives you two ways to find out everything you need to know about the latest (and past) updates: Google+ and release notes.

Get updates on Google+

View the latest information about Glass, including update features, on the Google+ Glass page at `http://plus.google.com/+GoogleGlass`. If you want to get new information about Glass features before they're announced to the general public, you can visit the MyGlass page (`www.google.com/myglass`) and read messages in the Explorer forums.

Read the release notes

Visit the Google Glass support page at `https://support.google.com/glass`. When you open the page, here's what to do:

1. **In the Exploring Glass section, click or tap Frequently Asked Questions.**

   A column of FAQ links appears on the right side of the page.

2. **In the FAQ links column, click or tap Release Notes.**

   The Release Notes page appears, displaying a link to the release notes for the latest version.

   You can also view archived release notes for earlier updates by clicking or tapping the Archive link and then clicking or tapping the release-notes link for the appropriate version.

Checking the OS version

Checking the version of the Glass OS that's installed on your device is easy. All you have to do is tap the Settings card in the timeline and then view the Device Info Settings card onscreen.

The Device Info Settings card (see Figure 13-1) tells you whether you have the latest OS, whether an update is available, and how much storage space is available; it also lists the serial number of your Glass.

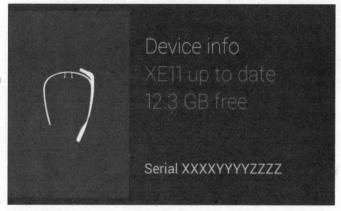

Figure 13-1: The Device Info Settings card tells you when the OS is up to date.

Device info
XE11 up to date
12.3 GB free

Serial XXXXYYYYZZZZ

If an update is available, you can prepare to charge your Glass and update the OS at the same time (refer to "Letting Glass Update Itself" earlier in this chapter).

Updating Glass Manually

If you know that the latest OS update is available and you want to get it right away, such as to fix a bug that's been. . . well, bugging you, you can. Here's how:

1. **Open the Settings bundle card on the timeline.**

2. **Tap the Device Info Settings card.**

3. Choose Update from the menu.

The update screen tells you that your Glass is updating (see Figure 13-2).

Figure 13-2:
An update in
progress.

Chapter 14

Hacking Your Glass

In This Chapter

▶ Giving users the best experience in your Glass app

▶ Deciding how to develop a Glass app

▶ Submitting your Glassware for inclusion on the MyGlass website

If you give a hacker a new toy, the first thing he'll do is take it apart to figure out how it works.

— Jamie Zawinski

When desktop computers were new and exciting (in the 1970s and 1980s), the term *hack* appeared in the computing lexicon. If you were a hacker who hacked a computer, it meant that you were able to figure out how to make the computer do useful, interesting, and/or unexpected things that you would share with others. Your hack inspired others to hack more computers and do more useful things. These days, however, the terms *hack* and *hacker* have negative connotations.

Glass brings new, exciting technology to the computing realm, so when we say *hacking* in this chapter, we mean teaching Glass new tricks, building something useful, and inspiring other Glass app developers.

Google wants as many people as possible to develop *Glassware* — apps for Glass — because (blunt honesty alert) if more Glassware is available, more people are likely to buy Glass. What's more, if you offer good-quality Glassware, your chances of submitting even more Glassware to the MyGlass website go up.

Giving Users What They Want

Glass has some specific user interface requirements that you need to follow, not only to create a great app, but also to give users the features they expect after they've used other Glass apps.

Don't just port an app written for a smartphone, tablet, or laptop to Glass, because the user interfaces for those other devices are fundamentally different from the Glass user interface.

Keep three main design principles at the top of your mind as you create your app:

✔ Ensure that the information on the screen doesn't distract the user from what's going on in the real world in front of him.

✔ Show information that's relevant and timely.

✔ Use an interface that integrates both visual and audial models.

Unobtrusive display

Glass is designed to provide quick information that adds to what the user is experiencing in real life. Too much information, however, subtracts from the user's experience. Users are looking at real life as well as information on the Glass screen, so it's important to make the information displayed on your Glassware as unobtrusive as possible.

One good example of how to use the Glass interface correctly is Google Search. If a user searches for *jellyfish* while viewing the creatures in an aquarium, for example, a brief definition of *jellyfish* appears on her Glass screen just above her line of sight (see Figure 14-1). The definition doesn't detract from her view of the jellyfish in the aquarium tank in front of her.

Figure 14-1:
A search
result
appears
slightly
above what
the user
sees in real
life.

What's more, information provided at inappropriate times will likely cause users to remove your Glassware. . .and tell all their friends to avoid it, too. Consider the example shown in Figure 14-2. The user wakes up very early in the morning and puts on his Glass to see what's going on. A few seconds later, he sees an ad for a cabbage sale.

Figure 14-2: A cabbage sale can be good or bad news, depending on your audience.

Chances are pretty good that most users don't need cabbage at 3:37 a.m., and most grocery stores are closed at that time anyway. If the user regularly shops at a 24-hour supermarket, however, and needs to get cabbage for the day's holiday meal right away, the ad may be appropriate. Your app can use the GPS connection between your app and your smartphone to "see" that the user is in or near the store and display information about that cabbage sale.

Relevant information

The information on the screen should be both current and relevant to what the user is seeing on her Glass so she's convinced that your Glassware is important. If the user isn't convinced, she won't let your wares take up valuable storage space on her Glass.

If a user needs to pop over to the grocery store to pick up some dinner, for example, she can use the Evernote app to display the shopping list on the Glass screen. This app saves the Glass user time because she doesn't have to fumble for her smartphone or a handwritten list to find out what's needed. Instead, she can focus on shopping.

Voice control

Users are going to use their voices to manipulate Glass and Glassware, so it's important to implement a visual and audial user model. After all, a user can start any app by using his voice, and there are plenty of examples about how to use voice control within an app. Figure 14-3, for example, shows the use of voice control in the Google Search app.

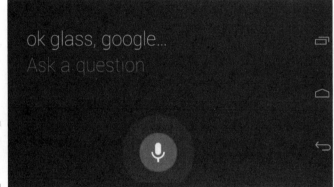

Figure 14-3:
The Google
Search app.

If you're going to include voice interaction in your Glassware, keep in mind the Glass rule that a voice command or message is initiated or sent *after* the user stops talking.

Choosing a Development Environment

You can use two types of development environments to build your Glassware: the Glass Development Kit (GDK) and the Mirror application programming interface (API).

Google produces the GDK so that you can build your Glassware directly on your Glass — useful if you want to use features such as location, sensors, and voice input. With GDK development, you can do basically the same things that you would on an Android smartphone or tablet, but your app needs to be designed to meet specific Glass user-interface requirements, discussed in "Giving Users What They Want" earlier in this chapter.

If you want to build web-based services that you can use to interact with Glass, use the Mirror API instead. This technology stores your Glassware code on the web so that you don't need to run the Glass app on Glass. Instead, your web-based service accesses your app code in the cloud and shows its information on the Glass screen.

If you develop a Glassware app with GDK and want to run that Glassware from a card on Glass, you can use the Mirror API to insert the card into the timeline. Then your users have it easy: They can just select the Glassware web app's card in the timeline and choose the app name from the menu.

Looking into the Mirror

Google recommends that you use the Mirror API to start developing your Glassware because you can write the code in a variety of programming languages, such as Java, PHP, and .NET.

Before you start, check out the Quick Start demo project to see how Mirror API code works (discussed later in this section) and get cues for building your own Glassware from that sample code. You can view the Quick Start project on your smartphone, tablet, or computer.

You need to have an existing web server so you can run your Glassware. You also need to know the address of the website where your Glassware project is stored so you can access that website from Glass and test your Glassware.

Start by visiting the Google Developers website to find Quick Start projects in each of the six programming languages that Google uses for the mobile API:

- ✔ **Go:** https://developers.google.com/glass/quickstart/go
- ✔ **Java:** https://developers.google.com/glass/quickstart/java
- ✔ **.NET:** https://developers.google.com/glass/quickstart/dotnet
- ✔ **PHP:** https://developers.google.com/glass/quickstart/php
- ✔ **Python:** https://developers.google.com/glass/quickstart/python
- ✔ **Ruby:** https://developers.google.com/glass/quickstart/ruby

Each website tells you how to access the Quick Start project for that language. For this example, we look at the Quick Start project that shows you how basic functions work in the Mirror API.

Here's how to find and run the project that tests the Mirror API functionality on Glass:

1. **Navigate to https://glass-java-starter-demo.appspot.com in your web browser.**

   An OAuth 2.0 permission request screen appears.

2. **Type the account username and password.**

   This account should be the same account that you use to access Glass.

3. **Click or tap the OK button.**

4. **Look at the controls of the project website, shown in Figure 14-4, and experiment with the features of the Mirror API.**

   The timeline items appear on the Glass screen.

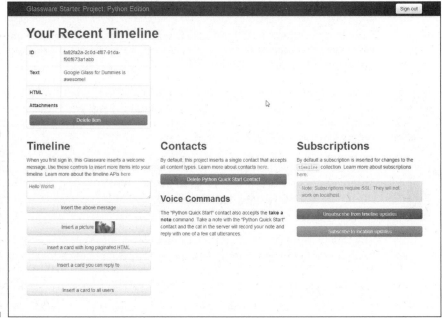

Figure 14-4:
The timeline section allows you to insert a text message, picture, or card into the Glass timeline.

Building from the kit

The Glass Development Kit (GDK) is for people who already know how to develop apps for the Android operating system and want to get building. Even if you're a seasoned Android developer, the differences in the Glass user interface create a learning curve.

Setting up the GDK

These instructions presume that you already have the ADT (Android Developer Tools) bundle installed and running on the computer you use to develop software. If you don't, download the ADT from the Android Developers website at `http://developer.android.com/sdk/installing/bundle.html`.

With that out of the way, here's how to set up GDK within the ADT bundle window:

1. **Click Window on the menu bar.**

2. **Click Android SDK Manager.**

   The Android SDK Manager window appears.

3. **In the Android 4.0.3 (API 15) section, check the boxes labeled SDK Platform and Glass Development Kit Sneak Peek, as shown in Figure 14-5.**

   The ADT bundle installs both items.

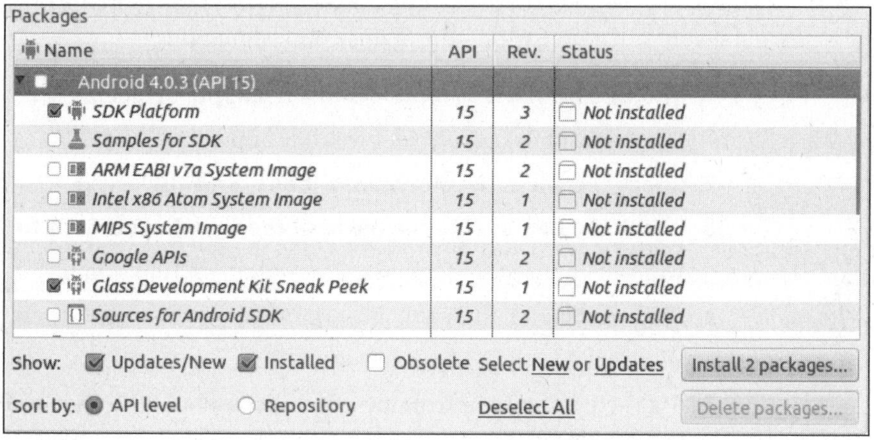

Figure 14-5: Select SDK Platform and GDK Sneak Peek.

Packages				
📱 Name	API	Rev.	Status	
▾ ☐ 🤖 Android 4.0.3 (API 15)				
☑ 📱 SDK Platform	15	3	Not installed	
☐ ⚒ Samples for SDK	15	2	Not installed	
☐ 🖿 ARM EABI v7a System Image	15	2	Not installed	
☐ 🖿 Intel x86 Atom System Image	15	1	Not installed	
☐ 🖿 MIPS System Image	15	1	Not installed	
☐ 📱 Google APIs	15	2	Not installed	
☑ 📱 Glass Development Kit Sneak Peek	15	1	Not installed	
☐ 🗋 Sources for Android SDK	15	2	Not installed	

Show: ☑ Updates/New ☑ Installed ☐ Obsolete Select <u>New</u> or <u>Updates</u> [Install 2 packages...]

Sort by: ◉ API level ○ Repository   <u>Deselect All</u>   [Delete packages...]

Connecting Glass to the GDK

To connect your Glass to the GDK, follow these steps:

1. **Put your Glass on your head, if you haven't done so already.**

2. **Select the Settings card.**

3. **Select the Device Info Settings card.**

4. **Tap Debug.**

When you're set up, you can verify the connection between your Glass and the GDK by opening the Eclipse app on your computer. This app allows you to create apps in a variety of languages, such as Java, Python, PHP, and C. You can confirm that Glass and the GDK are connected from within the Eclipse app by following these steps:

1. **Choose Window⇨Open Perspective⇨DDMS.**

   The DDMS window opens.

2. **Verify that Glass is listed on the Devices tab.**

Checking out sample Glassware

If you want to see examples of what Glassware looks like, the GDK contains several sample files for you to play with. Here's how to find and launch them on your computer within the Eclipse app window:

1. **Choose File⇨New Project⇨Android Sample Project.**

   The Android Sample Project window appears.

2. **Check the Glass Development Kit box, and click Next.**

3. **Check the box next to the name of the sample you want to run.**

   You can choose the Compass, Stopwatch, or Time sample.

4. **Click Finish.**

   You return to the Eclipse app window.

5. **Right-click the sample name in the list and choose Run As from the contextual menu.**

   A new window opens, asking whether you want to run the app on your computer within the Glass emulator or on the Glass device that's connected to your computer.

6. **Click Glass Device.**

   You see the sample Glassware running on the Glass screen.

Getting to work

When you've finished inspecting the sample files, put this book down and visit the Google Developers website so you can go through the interactive Building Your First App tutorial at `http://developer.android.com/training/basics/firstapp/index.html`. You can also explore the Developers website to see whether the other tutorials on the site strike your fancy.

You should build a few simple Glassware apps before you start building the Glassware of your dreams. Practice gives you the experience you need to prevent development headaches later.

Using only the Android SDK with Glass

The GDK gives you an all-access pass to Glass features. You can create Android apps by using just the Android Software Development Kit (SDK) on Glass, however, because Glass runs a version of the Android operating system. If you want to find out how to do this, and understand the benefits and pitfalls involved, you can find a good presentation on the Google Developers website at `https://developers.google.com/events/io/sessions/332704837`.

Submitting Your Glassware to Google

When you're satisfied that the Glassware you've developed is ready for users, submit it for inspection at the Glassware Distribution page (`https://developers.google.com/glass/distribute/index`).

Naturally, Google wants all the software it offers in MyGlass to be high-quality. To meet that goal, it requires developers to follow a launch checklist, which includes the following items:

✔ Review the app to ensure that it reflects Google's best practices for Glassware development. You can get the latest information about best practices at `https://developers.google.com/glass/design/best-practices`.

✔ Follow Google's branding and writing requirements, such as writing Glassware descriptions in English.

✔ If your site has a website that you use to promote your Glassware, follow Google's requirements for designing the site.

✔ Confirm that you tested the Glassware on your Glass by using the software as users would.

✔ Provide graphics and screen shots that can be used to promote your Glassware.

If you're not intimidated by this partial list, you can read the entire checklist of requirements at `https://developers.google.com/glass/distribute/checklist`.

As with any task that goes through a checklist review, the review process will go more quickly if you meet all these requirements to a T.

Checking out developer resources

If you want to get more information about developing on Glass and also connect with other developers, you can access some good resources online, including these:

✔ Tech maven Annie Gaus provides a good introduction to hacking your Glass at `www.youtube.com/watch?v=ANFu3Lvzil8`.

✔ For the latest information about designing on Glass by using both the Mirror API and the GDK, see `https://developers.google.com/glass/design/index`.

✔ For more details on Glass design, visit the Principles page of the Glass Developers website at `https://developers.google.com/glass/design/principles`.

Part V
Maintaining Your Glass

In this part . . .

✔ Know how to resolve common problems with Glass.

✔ See how you can keep your Glass in top condition as you use it in all sorts of weather.

✔ Find out how to get help with Glass.

✔ Discover how to demonstrate your Glass properly at `www.dummies.com/extras/googleglass`.

Chapter 15

Troubleshooting Common Glass Problems

· ·

In This Chapter
▶ Coping with common concerns
▶ Fitting Glass comfortably
▶ Fixing app and feature problems
▶ Going to Google for help

· ·

We can't solve problems by using the same kind of thinking we used when we created them.

— Albert Einstein

*T*hough it doesn't look like any other computer, Glass is still a computer, and that means you'll have some problems with it from time to time. Many of these problems are small and correctable, so you may want to bookmark this chapter for future reference.

Fixing Common Wonkiness

If your Glass is acting wonky (*wonky* is a technical term), here are some solutions to try for different situations.

Wi-Fi worries

If your Wi-Fi connection isn't working, you can forget it by following these steps:

1. **Tap the Settings card in the timeline.**

2. **In the Settings screen, open the Options menu by tapping the touchpad.**

3. **Choose Forget Connection from the menu.**

   Glass immediately forgets the connection so that you can add the Wi-Fi connection again or try another one.

For details on Wi-Fi setup, see Chapter 5.

Tethering troubles

If you're using Glass with your Android smartphone or iPhone to connect to a cellphone provider's Internet network, you know that when your Wi-Fi connection isn't available, you can use your tethered data connection to the provider's network. (That is, you can if you're in range of a cellphone tower's signal.)

If your Glass says that you're connected to the Internet but you can't do anything Internet-related, such as browse the web or check your e-mail inbox, you can forget the data connection in much the same way that you can a Wi-Fi connection. Then you can restart the data connection and see whether your Glass can use the Internet.

On an Android smartphone

If you have an Android smartphone, here's how to forget the data connection and restart it:

1. **Tap the Settings card in the timeline.**

2. **Tap the More Settings card.**

3. **Tap Mobile Networks.**

4. **Turn off your data connection by clearing the check box in the Data Enabled section.**

5. **Wait about 20 seconds and then reestablish the data connection by checking the Data Enabled check box.**

On an iPhone

If you have an iPhone, here's how to turn the data connection off and on again:

1. **Open the Settings screen.**

2. **Tap Cellular.**

3. **Swipe the Cellular Data button from right to left.**

4. **Wait about 20 seconds and then reestablish the data connection by swiping the Cellular Data button from left to right.**

If you're using another computing device that has a working Internet connection — such as a tablet, laptop, or desktop computer — you can also check for documentation on the device provider's website.

Bluetooth bummers

If your Glass is telling you that you made that left turn in Albuquerque and now you're in Los Angeles when you're actually in a bullfighting ring, the Bluetooth connection with your smartphone isn't working correctly. You can apply any of three potential remedies:

✓ On your smartphone, turn Bluetooth off. When Glass tells you that the Bluetooth connection is forgotten, you can reestablish the connection with your smartphone (see Chapter 6).

✓ Turn off your smartphone and then turn it back on. When you turn off the phone, you immediately lose the Bluetooth connection, but the smartphone will find the Glass Bluetooth connection after you turn it back on.

✓ If you're using the MyGlass app on your smartphone (see Chapter 5), quit the app and then restart it.

General glitches

One good fix for any misbehaving computing device is to close it down and restart it. Glass is no different. You can hard-reset the device by holding your finger on the power button for 15 seconds, no matter what you're viewing onscreen. After 15 seconds, you hear a higher-to-lower chime. Release the touchpad. When Glass turns off, wait a few seconds, turn Glass back on, and see whether the problem is solved.

If not, the last resort is to perform a factory reset that restores Glass to its original factory settings. Here's how:

1. **Tap the Settings card in the timeline.**

2. **Tap the Device Info Settings card.**

3. **Choose Factory Reset from the menu.**

   A warning onscreen tells you that you'll lose all data unless you've backed up that data to an external source such as Google+ Auto Backup (see Chapter 8).

If you want to save any data stored on your Glass, such as pictures and videos, be sure to check your Google+ Auto Backup account on another device, such as your smartphone or computer, to confirm the photos and videos are there.

Begin the restoration process by selecting Yes on the screen. After your device's factory settings are restored, you can start setting up your accounts, connections, and apps. To find out how to set up Glass, see Chapter 3.

Fitting Glass to Your Face

As with any pair of glasses, you may have trouble wearing Glass at first. You may also find damage associated with normal wear and tear. Here are some ideas for solving various problems related to putting Glass on your head.

Finding the Goldilocks fit

If you wear eyeglasses, chances are that you'll have a good idea how to fit Glass on your head comfortably because you're used to adjusting your eyeglasses. But if you can't get the right fit no matter what you try, or if you're not used to wearing glasses, consider trying one or more of the following solutions:

- ✔ If you're picking up your Glass at a Google Base Camp in New York, Los Angeles, or San Francisco, find out whether you can get fitted there after you purchase the device online. Also check to see whether you can get refitted at that Base Camp later, if necessary.

- ✔ Visit the Google Glass Help site or call a Glass Guide for live help (see "Getting Help from Google" later in this chapter).

- ✔ Consider getting a new frame for your Glass that may provide a more comfortable fit. We cover changing frames in Chapter 11.

Getting accustomed to your Glass

Your eyes and your brain aren't used to having a screen right in front of them, so you may experience pain in your head — not only from an ill-positioned Glass, but also from eyestrain as you try to get used to the Glass experience.

Using Glass is like exercising. When you start an exercise program to get fit, you don't run a marathon right away. Likewise, you don't use Glass for hours at a time right away. Instead, use your Glass for short periods at first, adding a few minutes each day. Any time you feel uncomfortable, take Glass off, and wear it again when you feel like it.

Fixing a crack in the frame

Glass is a pretty sturdy device, but like any pair of glasses, it's subject to wear and tear over time. One common signal of wear is a crack in the frame, especially near joints such as those between the frame and nose pads.

When you find a crack in the frame, here are some solutions you can try:

- ✔ Visit the Glass Community (`https://www.glass-community.com`) to find out what solutions other users have employed. You may also want to take a picture of the crack with your smartphone and post it on the site to see whether any users and/or Glass Guides can provide suggestions.

- ✔ Contact a Glass Guide by phone or e-mail to get some ideas. The Guide may want to start a Google Hangout on your computer or smartphone so he or she can see exactly what the problem is and offer solutions. (For more on Hangouts, see Chapter 9.)

- ✔ If all else fails, the Guide may arrange to have Google ship you a replacement frame.

For more information, see "Getting Help from Google" later in this chapter.

Resolving Operation Errors

If you find that certain functions aren't working on Glass as described in this book, here are some ideas for resolving problems with different features and apps.

I see dead pixels

Google works hard to ensure that all the pixels on each Glass function properly when the device comes out of the factory. As with any technology, however, problems happen, and you may be the unlucky owner of a Glass that has one or more dead pixels. *Dead,* in this context, means that the pixels are black or unlit no matter what comes up onscreen. At best, dead pixels are really annoying; at worst, they obscure important information.

If you see dead pixels, don't hesitate to visit the Google Glass Help site and contact a Glass Guide (see "Getting Help from Google" later in this chapter). Google offers a 30-day warranty, so if your Glass is damaged or not working correctly, move quickly to get a replacement device.

My map is frozen

Sometimes, an app stalls after an upgrade. You may see stalling issues with the Maps app after an update, for example, because of Bluetooth synchronization issues between your Glass and your smartphone.

The solution is straightforward: On your smartphone, turn off Bluetooth. When Glass tells you that the Bluetooth connection is forgotten, reestablish the Bluetooth connection on your smartphone.

Callers sound like they're underwater

Glass uses bone-conduction transducer (BCT) technology in its speaker. BCT picks up the vibrations from your skull to help improve call quality. You may have heard of BCT because it's used in smartphones, earbuds, and head-phones as well.

Despite this technology, you may still have trouble hearing people on the phone, especially in noisy areas. You can take some steps to fix this problem:

- ✔ **Reposition the speaker.** Check the position of the speaker to ensure that it comes up right against your ear. The speaker works best when it's right up against your ear.

- ✔ **Turn up the volume.** If the voice you're listening to is too soft, tap the Settings card in the timeline and then tap the Volume Settings card to view and change the volume.

  This procedure may seem to be elementary, but when you're in a loud place and your caller's voice is suddenly too soft, you may not think to turn up the volume.

- ✔ **Connect the earbuds that came with your Glass to the device.** The ear-buds include printed instructions that tell you how to connect them to Glass.

- ✔ **Move.** If you still can't hear the person adequately with either the speaker or the earbud, consider placing your hand over your free ear and/or moving to a quieter location.

If all else fails, ask a Glass Guide to talk you through more troubleshooting steps (see "Getting Help from Google" later in this chapter).

I have files and I can't back up

If you can't use the Auto Backup feature to back up one or more files from your Glass to your Google+ Auto Backup account, here are some ideas:

✔ Turn off your Glass, turn it back on, and see whether Auto Backup works the way you expect.

✔ If you have a large number of files to back up, especially pictures and videos that take up a lot of space, these files back up automatically when you charge your Glass while the device has an active Wi-Fi connection.

When you try to back up while Glass is running on battery power, the backup process goes much more slowly, and the battery may run out of power before the backup is completed.

✔ Check your Google+ account settings to ensure that you have enough space in your Auto Backup account to accommodate all the files you want to back up. The default amount of space in your backup account is 15GB. You should receive a notification from Google when you're running low on space, but if you don't (or if you're just curious about how much room you have left), you can find out how to pay for 100GB and 200GB plans at `https://www.google.com/settings/storage`.

✔ Connect your Glass to your computer and then sync your pictures, videos, and other files to your computer (see the next section).

Pictures and video won't sync to my computer

When you connect your computer to your Glass by using the micro USB cable that's included with Glass, your computer should recognize Glass as another external storage device and display it in your list of storage devices. Though the days of "Plug 'n' Pray" are over, there may be times when your computer can't find Glass when it's connected to your computer. If you have this problem, here are some solutions to try:

✔ Remove the micro USB cable from the computer, wait a few seconds, and then connect the cable to a different USB port on the computer.

✔ Turn off Glass, remove the micro USB cable from the computer, turn Glass back on, and then connect the cable to the same or a different USB port on the computer.

✔ If you can't use your computer's file-management app (such as File Explorer on a Windows PC or Finder on a Mac), try using another app to move the files.

You can use Google's Picasa image-editing software, for example. It's web-based, it recognizes Glass automatically when you have the device connected to your computer with a micro USB cable, and it allows you to move your photos between your Glass and your computer.

If none of these solutions solves your problem, you may have a defective micro USB cable. Contact a Glass Guide (see the next section) to find out whether the cable needs to be replaced.

Getting Help from Google

We hope that the steps we suggest in this chapter will get your Glass working the way you expect and let you get on with your life. If our suggestions don't work, however, or if you don't see your problem listed here, you can get help on the Google Glass Help site at `https://www.google.com/glass/help/#get-started` (see Figure 15-1).

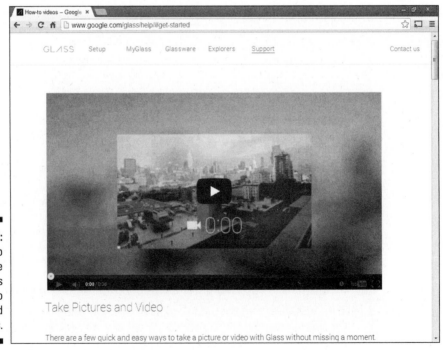

Figure 15-1:
The help website includes how-to videos and articles.

On this website, you can find tutorials; check for potential solutions to your problems; talk to other users in the Glass Community; or get live, one-on-one help for free from a Glass Guide by calling (800) 452-7793 (a toll-free number) from 5 a.m. to 9 p.m. Pacific time.

If you prefer to send an e-mail message to a Glass Guide, use the form on the help website or e-mail `glass-support@google.com`. A Guide will make every effort to reply within 48 hours.

Chapter 16

Keeping Your Glass in Top Condition

Success is a science; if you have the conditions, you get the result.

— Oscar Wilde

Glass is a rugged piece of technology that can handle the rigors of day-to-day use. Over and over every day, you put Glass on and take it off, press your finger to the touchpad, and use the camera to take plenty of pictures and videos.

As with any other piece of technology, you need to know how to treat your device kindly so you can enjoy it for years to come. For one thing, you need to protect it from various hazards. The frame (and perhaps more) will break if you forget to take Glass off your couch before your large indoor dog sits on it. (And the dog *will* sit on it.)

You also need to know how to store Glass properly, purchase genuine Glass replacement parts, keep the device fully charged, and keep the operating system updated so you can get the most out of your device. We show you how in this chapter.

Handling Glass with Care

As with any other product, the performance of your Glass depends on how well you take care of it. Glass has several sensitive parts that you must treat with care. Fortunately, the device comes with its own pouch so you can store your Glass safely when you're finished using it.

Keep it right side up

Don't take off Glass and leave it upside down. Why, you ask? You may think that Glass is like any other pair of glasses. It's not. For one thing, it's heavier than a regular pair of glasses. Also, positioning is important. When you put your Glass on, the camera and screen unit are on the right side of the device. If you place the device upside down, as shown in Figure 16-1, especially on a hard surface, its weight will press the Camera button — and Glass will take pictures and record video even if it's turned off. The next time you put on your Glass, you could find that it's storing a lot of pictures or a long video that you didn't expect. Worse, taking those pictures and videos may have drained your battery completely.

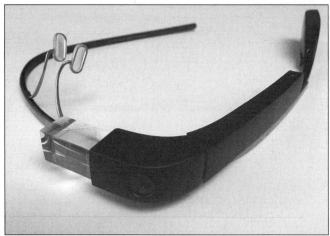

Figure 16-1:
When you rest your Glass upside down, the device presses down on the Camera button.

What's more, if you don't have a Glass shield and constantly leave your Glass on a hard surface, the nose pads will move slightly, and in time, the device won't feel comfortable anymore. (For details on where to get a shield, see Chapter 11.)

If you connect the micro USB cable that came with your Glass to your charger, Glass will rest the right way while you're charging it.

Watch the weather forecast

Glass is sensitive to the elements, so don't take it out in the rain or on a humid day (especially in the Southern or Midwestern United States in the summer).

Even if no actual rain is falling, or if you put a hoodie on over your Glass, water can condense within the unit. Then you may find that Glass isn't working right. It may just shut itself off, for example. Putting Glass in dry rice won't solve the problem.

What's more, the reflective foil around the screen cube may ripple or possibly tear in high humidity. If the foil ripples, you're going to have a much harder time seeing the Glass screen; if the foil tears, you won't be able to see the screen at all.

At this writing, Google hasn't published temperature limits for Glass, but you can always get the latest tech specs for the device at `https://support.google.com/glass/answer/3064128?hl=en&ref_topic=3063354`.

Store it in a pouch

Glass comes with a pouch (see Figure 16-2) that protects it from scratches, dust, and the elements. The bottom of the pouch contains a hard protective shell, so if you drop the pouch, your Glass will still be protected. The pouch stores Glass safely with or without a protective shield (see Chapter 11).

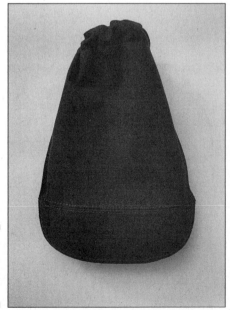

Figure 16-2:
A closed
Glass
pouch.

If you misplace your pouch or want to get an extra for use at another location (one for home and one for the office, for example), you can purchase another pouch from the Glass Store at `https://glass.google.com/getglass`.

Don't panic if you don't find a pouch in the Glass Store; the item may be temporarily out of stock. Check back another day.

Empowering Your Glass

You should always keep your Glass charged and updated, even if you're not going to be using it. Otherwise, you may miss interesting moments because your battery has run out of power or you haven't installed an upgrade.

When you took your Glass out of the box for the first time, you probably saw that it came with a charger unit and a micro USB cable. The cable and charger are two-tone so that you know the correct way to place the cable. The micro USB port on your Glass is white on the bottom, so you can match the white at the bottom of the port to the white portion of the cable.

On your cable (see Figure 16-3), one side of the port is white, which tells you to insert the white side of the micro USB cable into the white side of the power port on your Glass; the white side of the cable appears on the outside of your Glass.

Figure 16-3:
One side of the cable port is white, to plug into the charger; the other side plugs into the Glass device.

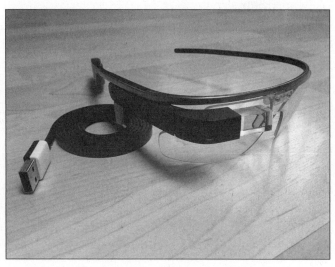

You'll know that it's time to charge Glass when you see the empty-battery icon on the Glass screen (see Figure 16-4). As Glass charges to full power, the LED near the power button pulsates slowly, and the battery icon updates the charge-percentage status. When the screen says the charge is at 100 percent, the battery is fully charged, and Glass is ready for you to unplug and use.

Figure 16-4:
The empty-battery icon tells you that Glass needs to be charged.

You can also charge Glass (as well as transfer files) by plugging the micro USB cable into an available USB port on your computer.

Whether you charge Glass by using the charger or your computer, if you're in range of a Wi-Fi network, Glass automatically updates its operating system and synchronizes new photos and videos with your Google+ Auto Backup account. For details on updating the operating system, see Chapter 13.

If you misplace your charger or need a charger for a separate location, you can purchase another charger from the Glass Store at `https://glass.google.com/getglass`.

Always get a genuine Google Glass charger so you can be assured that you won't be shocked by power or charging problems.

Chapter 17

Getting Help with Glass

Tell everyone what you want to do and someone will want to help you do it.

— W. Clement Stone

*Y*ou may have turned to this chapter because you're having a problem with your Glass and none of the suggestions in other chapters in this book has solved it, or you may just want to familiarize yourself with the support options you have available.

Google offers plenty of help options, as you see in this chapter.

Checking Out the FAQs

The Google Glass Help website contains several articles that answer frequently asked questions (FAQs). You can view the FAQs in various categories at https://support.google.com/glass. The Help page, shown in Figure 17-1, displays the categories. By default, the Get Started category is open, listing the topics within the category.

You can view topics within a category by clicking or tapping the category name. View more information about a topic by clicking or tapping its name in the list; an article on that topic appears (see Figure 17-2) so you can find out whether the information it contains answers your question. In some articles, a Next Steps section appears at the end; click or tap the links in that section to read articles that contain related information.

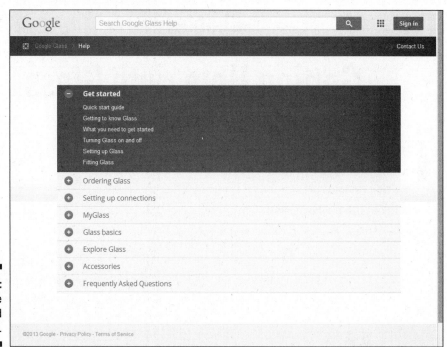

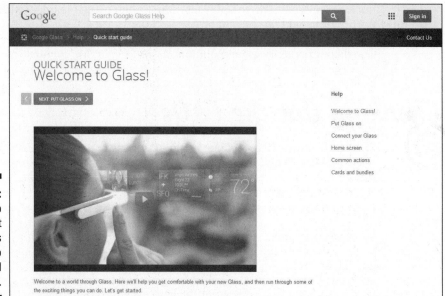

To the right of each help article is a list of links to related articles. Above the article is a gray bar, the left end of which contains a breadcrumb-style navigation system that shows where you are in the help system. The page you're on appears as a highlighted link at the right end of the breadcrumb list (Help, in Figure 17-2), and parent page links appear to the left of the current page (Google Glass, in Figure 17-1). Go to the page by clicking or tapping the link in the breadcrumb list.

You can also search for a particular topic by typing the character(s) and/or word(s) in the Search Google Glass Help box at the top of any page on the help website. After you type the character(s) or word(s), begin searching by clicking or tapping the blue search button to the right of the search box.

Going to Google for Help

If you can't find the answer to your question on the Google Glass Help website, you have two ways to contact a Glass Guide by phone: Call Google directly, or visit the help website and ask a Glass Guide to call you.

Calling Google

Call Google at (800) GLASSXE — which is (800) 452-7793 if you don't like to hunt and peck on your telephone keypad. Glass Guides are available by phone every day from 5 a.m. to 9 p.m. Pacific time. The only exception to this rule is during the winter holiday season from December 20 through January 4. Check the Google Glass Help website at `https://support.google.com/glass/answer/3079854` to get the latest news about phone availability.

When you purchased your Glass, you received your unique Glass ID in an e-mail message sent to the address you provided Google. You need to have this Glass ID at the ready so that the Glass Guide can identify you, call up your Glass information, and help you more effectively.

Requesting a call from a Glass Guide

One of life's axioms is that when you talk to technical support, you have to wait. Google gets around this problem by offering to have a Glass Guide call you when one is available. You need to log into your Google account and then visit the Google Glass Help website to get started. Next, follow these steps:

1. **Click or tap the Contact Us link in the top-right corner of the Support home page.**

   The Contact Us page appears, as shown in Figure 17-3.

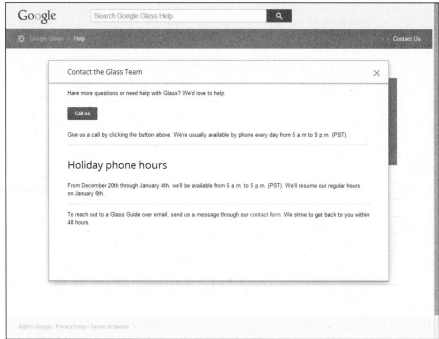

Figure 17-3:
The Contact
Us page
helps you
get in touch
with Google.

2. **Click or tap the Call Us button.**

3. **Type your name, e-mail address, and phone number in the appropriate fields.**

   The estimated time before a Guide will call you appears below the phone number (see Figure 17-4).

4. **Click or tap the Call Me button.**

   A new screen appears, thanking you for submitting the form and telling you that a Glass Guide will call you shortly.

The estimated time for a Glass Guide to call you is just an estimate. It may take less or more time for a Guide to call. Be sure to stay near your phone (or take your phone with you) and have the ringer turned on.

Asking for help by e-mail

If your problem isn't a pressing one, and you prefer to talk with a Glass Guide online instead of on the phone, Google gives you two options. The first option is to send an e-mail message to glass-support@google.com. The other option is to fill out a contact form on the Google Glass Help website, like so:

1. **Visit http://support.google.com/glass.**

2. **Click or tap the Contact Us link in the top-right corner of the page.**

3. **In the resulting page, click or tap the Contact Form link (see Figure 17-5).**

   The contact form opens.

4. **Fill out the contact form (see Figure 17-6).**

   Items with an asterisk after them are required; you must fill in these boxes for Google to accept the form. If you don't fill in a required box, the form reappears, listing the boxes you need to fill in.

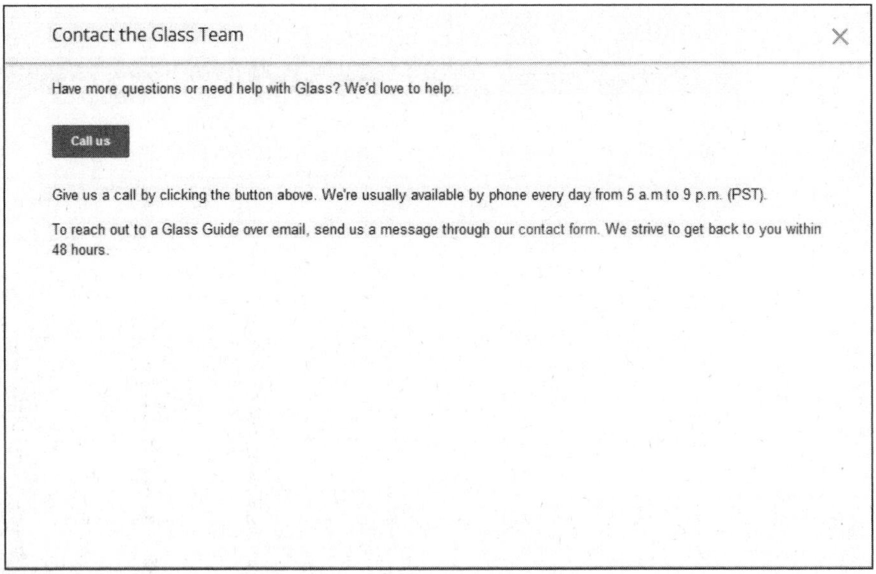

Contact the Glass Team   ✕

Have more questions or need help with Glass? We'd love to help.

Call us

Give us a call by clicking the button above. We're usually available by phone every day from 5 a.m to 9 p.m. (PST).

To reach out to a Glass Guide over email, send us a message through our contact form. We strive to get back to you within 48 hours.

Figure 17-5:
Click the
Contact
Form link.

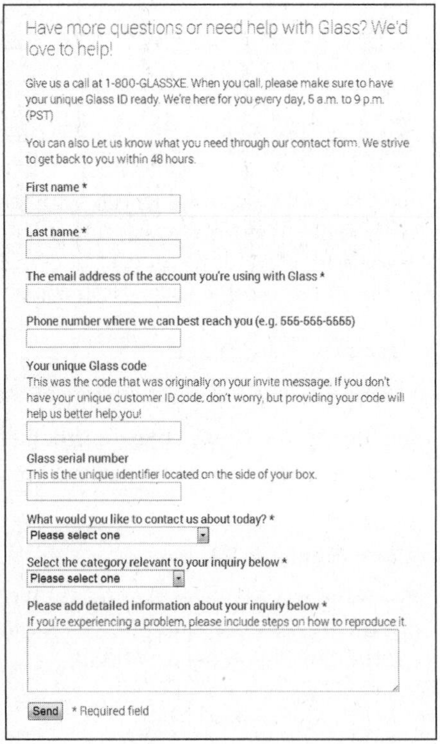

Have more questions or need help with Glass? We'd love to help!

Give us a call at 1-800-GLASSXE. When you call, please make sure to have your unique Glass ID ready. We're here for you every day, 5 a.m. to 9 p.m. (PST)

You can also Let us know what you need through our contact form. We strive to get back to you within 48 hours.

First name *

Last name *

The email address of the account you're using with Glass *

Phone number where we can best reach you (e.g. 555-555-5555)

Your unique Glass code
This was the code that was originally on your invite message. If you don't have your unique customer ID code, don't worry, but providing your code will help us better help you!

Glass serial number
This is the unique identifier located on the side of your box.

What would you like to contact us about today? *
Please select one

Select the category relevant to your inquiry below *
Please select one

Please add detailed information about your inquiry below *
If you're experiencing a problem, please include steps on how to reproduce it.

Send   * Required field

Figure 17-6:
Provide
as much
information
as possible
to help the
Glass Guide
solve your
problem.

5. **Click or tap the Send button.**

   A confirmation message appears, thanking you for your submission and promising that a Glass Guide will respond as soon as possible.

Google strives to reply to your e-mail message within 48 hours of receipt, but that's not a guarantee. If you find that Google hasn't responded to your request for help, you can always send a follow-up message or call Google for immediate help.

If you're close to the major U.S. metropolitan areas of New York, Los Angeles, or San Francisco, you can visit Glass Guides at Google Base Camps in those cities.

Turning to the Glass User Community

You may be able to get information about any problems you have from other Glass users and from Google itself. The Glass Community website is the place to exchange messages with other Glass users if you have a problem or you want to discuss experiences. Google also maintains a Google Glass profile on Google+ so you can get the latest information and operating-system updates for Glass.

Glass Community

You can send messages to and receive messages from other Glass users at the online Glass Community, available at `https://www.glass-community.com`. This community is sponsored by Google and features messages in a variety of categories.

The community home page, shown in Figure 17-7, is divided into two sections:

- ✔ The left side of the screen contains several tiles of information you may be interested in. The Top Kudoed Posts tile, for example, represents the posts that the most people are talking about. You can view the information contained within a tile by clicking or tapping the button in the bottom-right corner of the tile.

- ✔ The right side of the screen displays your account information, a search box, a list of message-board topics, and the number of messages on each topic. To view messages, click or tap a topic name.

Below the list of boards is a list of the latest posts from Google+ about Glass. We discuss Google+ in the next section.

Google+

To see all the latest Glass updates on Google+, visit `https://google.com/+GoogleGlass`. Like the user forum discussed in the preceding section, the Google Glass page on Google+ (see Figure 17-8) contains tiles representing posts about Glass, such as the latest announcements and software-update information.

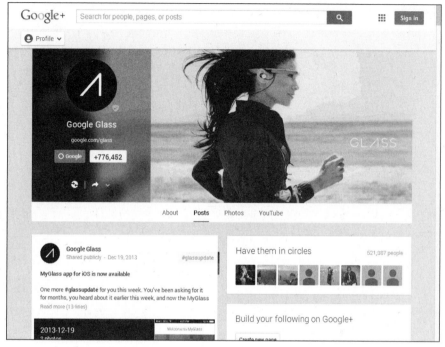

You can get information from this page in several ways:

- ✔ Follow the page in your Google+ feed by tapping the Follow button.
- ✔ Tap About to get more information about the page and links to related sites, such as the Glass page on YouTube.
- ✔ View photos of Glass in action and on people's heads by tapping Photos.
- ✔ Watch various how-to and informational videos about Glass by tapping YouTube.

To return to the posts page, tap Posts. You can view information within a post by tapping the post's tile.

Part VI
The Part of Tens

Enjoy an additional Part of Tens chapter online at www.dummies.com/extras/googleglass.

In this part . . .

✔ Discover apps that make your daily life with Glass more productive and satisfying.

✔ Know the ten things you shouldn't do with Glass.

✔ Find out how to connect with other Glass users and enthusiasts online and in person.

✔ Enjoy an additional Part of Tens chapter online at www.dummies.com/extras/googleglass.

Chapter 18

Ten Apps You Need to "See"

Man's greatness consists in his ability to do and the proper application of his powers to things needed to be done.

— Frederick Douglass

At the time this book was written, fewer than 100 apps were available for Glass, but the fact that even that many apps were available for a project still in the development and testing stage attests to the excitement generated by this new type of computing device. By now, you may have hundreds or even thousands of apps to choose among.

Even so, some of the early apps were developed for a reason: They take great advantage of Glass to show you the things can do to make your life better. Google has made several free apps available at `https://glass.google.com/glassware`.

The apps in the Glass Store are approved for use with Glass, but those aren't the only apps you can use. One good resource where you can view third-party apps, as well as the latest Glass updates, is the GlassAppSource website at `www.glassappsource.com`.

Google Apps

Because Glass is a Google product, it makes sense that Google made some of its most popular apps available for Glass on the day the device was released.

Hangouts

Hangouts (see Chapter 9) is Google's instant-messaging (IM) app that allows you to connect in a live text or video chat with anyone who's using the Hangouts app on a computer, tablet, or smartphone.

You can send photos to other Hangouts members while you're on a text or video chat, and when you're on a video chat, other people can see what you're viewing on your Glass.

You should download Hangouts, not only because this app is a great use of Glass, but also because Glass Guides (which is what Google calls Glass support representatives) may ask you to engage in a Hangout if you have a problem.

Google Play Music

Google has its own music catalog on the Google Play website, so it probably comes as no surprise that Google created an app for Glass that allows you to access the Play Music website directly without having to visit the Google Play site.

You can use the Play Music app to find songs and albums, purchase songs and albums, and download them to Glass so you can play songs whenever you want without having to fumble for your smartphone and/or headphones.

Google makes it easy to use the Glass interface to navigate the Play Music catalog. You can search by song, album, or artist. You can also create a playlist of your favorite songs so you can play one or more songs within that playlist.

You don't even have to use the touchpad to listen to a song. While you're viewing the timeline screen, you can just say "OK Glass, listen to. . ." followed by the name of the song, album, playlist, or all songs by the artist, as shown in Figure 18-1.

When you listen to an album or playlist, the app starts playing the first song in the album or playlist. If you listen to all songs by an artist, the app finds the first song by the artist you downloaded and starts to play that song.

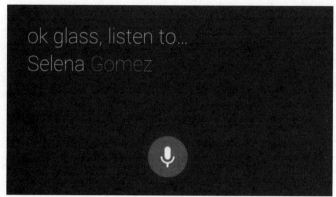

Figure 18-1:
Start lis-
tening to
a song
by saying
"OK Glass,
listen to."

YouTube

Google purchased the video-sharing website YouTube in 2006, and YouTube is easily one of the most popular websites today. Because taking videos is a popular use of Glass, Mr. Spock would agree that it's logical that Google created a YouTube app for Glass so you can share the videos you take on Glass easily.

Here's a quick primer on how to publish your videos by using the YouTube app for Glass:

1. **Record the video or select the card in the timeline that contains the video you want to share.**

2. **Tap the touchpad and then choose Share from the menu.**

3. **Choose the channel to which you want to publish your video.**

   You can choose the default channel you created when you set up your YouTube account, or you can select another channel.

4. **Choose the privacy setting you want to use (see Figure 18-2).**

   You can choose Public (all YouTube users can watch the video), Unlisted (only you can watch it), or Private (only invited YouTube users can watch it).

   The YouTube app publishes your Glass video to the channel automatically. When the app finishes publishing the video, it plays the uploaded video so you can verify that it looks as you expect.

Figure 18-2:
Decide how
public you
want your
video to be.

You can also verify that your video uploaded correctly by opening YouTube on another device, such as your computer, and opening the appropriate channel to see whether the video is there.

Social and Connectivity Apps

Some interesting apps let you use Glass to keep apprised of what's happening, communicate with people who don't speak your language, and access your Facebook account.

IFTTT

IFTTT (an acronym for *If This Then That*) lets you create connections between apps by saying "If this, then that." The *this* part of the argument is a trigger such as "I'm tagged in a Facebook photo." The *that* part of the argument is an action such as "Send me an e-mail message." The app calls this completed argument a *recipe* that you can save and activate so that Glass will remind you when the trigger event occurs.

Several premade recipes are available, based in part on your current location. Figure 18-3, for example, shows weather recipes for a specific location.

If you want the IFTTT app to find your current location, you need to set up the MyGlass app on your smartphone (see Chapter 5) so that the app can find you by using the smartphone's GPS feature.

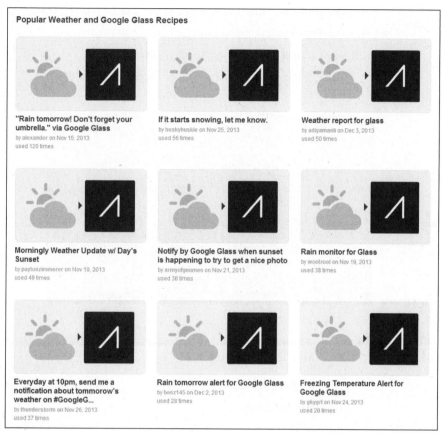

Popular Weather and Google Glass Recipes

"Rain tomorrow! Don't forget your umbrella." via Google Glass
by alexander on Nov 15, 2013
used 120 times

If it starts snowing, let me know.
by huskyhuskie on Nov 25, 2013
used 56 times

Weather report for glass
by adiyamanli on Dec 3, 2013
used 50 times

Morningly Weather Update w/ Day's Sunset
by paytonzimmerer on Nov 19, 2013
used 49 times

Notify by Google Glass when sunset is happening to try to get a nice photo
by armyofgnomes on Nov 21, 2013
used 38 times

Rain monitor for Glass
by wootroot on Nov 19, 2013
used 38 times

Everyday at 10pm, send me a notification about tommorow's weather on #GoogleG...
by thunderstorm on Nov 26, 2013
used 37 times

Rain tomorrow alert for Google Glass
by benz145 on Dec 2, 2013
used 28 times

Freezing Temperature Alert for Google Glass
by gkygrl on Nov 24, 2013
used 28 times

Figure 18-3:
Weather recipes in IFTTT.

Facebook

Facebook is as close to a ubiquitous social networking website as you can get. You don't need to check your phone for Facebook updates anymore, because now you can get that information directly on your Glass.

The first thing you need to do is download and install the Facebook Glassware from the MyGlass website or get the smartphone app. After you're in MyGlass, here's what to do:

1. **Activate the Facebook app by clicking or tapping the Facebook card.**

2. **If you're not logged in to Facebook, type your Facebook username and password in the appropriate fields and then click or tap the Allow Access button.**

3. On the Glass timeline screen, select the Facebook card.

The Facebook posts appear as cards.

4. View a post by tapping the appropriate card.

The post appears onscreen, as shown in Figure 18-4.

Figure 18-4:
A Facebook post with a photo.

It's easy to share photos and videos you've taken on Glass with your Facebook friends. Here's how:

1. Take a photo or video on Glass.

2. Tap the touchpad.

3. Choose Share from the menu.

4. Choose a sharing option: Only Me, Friends, or Public (that is, every-one on Facebook).

You can verify that your photo or video was posted to Facebook in the Facebook Glass app.

You can also use Facebook to share a photo or video stored on Glass by finding the photo or video card on the timeline and then repeating Steps 2 through 4.

To find out how to take photos and videos with Glass, see Chapter 8.

Useful Utilities

Several Glass utility apps help make your life a little bit easier. Here are a couple of examples for you to consider.

Word Lens

The Word Lens app is another cool example of what you can do with Glass. If you have text that's printed in another language, you can translate it into your preferred language, as shown in Figure 18-5. (By the way, the text in the figure means *Welcome to the future.*) All you have to do is hold your head still while you look at the printed page; after a few seconds, you see the translation in the top-right corner of the screen.

Figure 18-5: Translate onscreen text into your preferred language.

The text needs to be as clear as possible for the Word Lens app to translate the text successfully.

You can get the best translation results when the text is printed on a laser printer, in an easily recognizable font such as Arial, and at a fairly large size.

The app is self-contained, so you can translate text without using an active Wi-Fi connection to the Internet.

You can translate text between English and the following languages:

- ✔ French
- ✔ German
- ✔ Italian
- ✔ Portuguese
- ✔ Spanish

Stopwatch

If you find that you need to time people on a certain task, such as completing a test, you don't need to pull out your smartphone and keep checking the time periodically. Instead, you can install the Glass Stopwatch app and keep the timer in front of your face so you can be ready to take action as soon as time's up.

Here's a how to use the Stopwatch app after you've installed it on Glass:

1. **Tap the touchpad and then choose Set Timer from the menu.**

2. **Select the timer measurement: Hours, Minutes, or Seconds.**

3. **Set the number of hours, minutes, or seconds the timer will run, depending on the measurement type you set in Step 2.**

4. **Start the stopwatch by choosing Start from the menu, as shown in Figure 18-6.**

   You can work on other tasks on Glass while the timer is running. When the timer reaches zero, the speaker sounds an alarm.

5. **Go back to the Home screen; swipe to the left, tap the Timer card, and then choose Stop from the menu.**

Figure 18-6: Start the timer by choosing Start.

Weather Alert

When you're out and about, you may want to know when bad weather is coming your way. Even on a clear day, you may need to check the air quality in your area. The Weather Alert app from Roll Innovation (`https://glass.whyroll.com/weatheralert/info`) is a good app for getting up-to-date information right on the Glass screen as alerts are broadcast.

The Weather Alert app uses National Oceanic and Atmospheric Administration (NOAA) data, so you can use the app only in U.S. states and territories. Also, not all areas of the United States are monitored, so contact Roll Innovation to get the latest coverage map for your area.

When a weather alert is issued, you see the alert card on the timeline screen, showing the type of alert, a map of the affected area, and the duration of the alert. This card presents only a summary of information, so if you want to listen to the entire alert as broadcast on NOAA weather radio, tap the touchpad and then choose Read Aloud from the menu. You hear the report through the Glass speaker.

Games and Leisure Apps

Google believes that you can play as hard as you work on Glass, so a good many apps allow you to enjoy your hobbies and down time. Following are a couple of leisure apps that you can download and check out to see whether they're right for you.

Allthecooks

The Allthecooks app is almost like having a variety of cooks in the kitchen with you. The app gives you access to more than 150,000 recipes that have been added to its database by Allthecooks personnel as well as other Allthecooks members. You can contact those members to get feedback about recipes and ask questions.

When the app is running, you can say "OK Glass, find a recipe for. . ." and then the name of the dish you want, such as pizza (see Figure 18-7). When you access a recipe on Allthecooks, the recipe appears on the Glass screen in step-by-step format so you can prepare the recipe without having to use your dusty and/or greasy fingers to flip pages in a cookbook.

Figure 18-7:
You can search for a specific dish to make.

The first step of each recipe tells you how long it takes to prepare, how many people it serves, and how much of each ingredient you need. You can move to the next step by swiping forward on the touchpad. If you want to go to the previous step, swipe backward on the touchpad.

Spellista

If you enjoy solving puzzles and/or want to give your kids a creative and fun way to learn to spell, consider downloading the Spellista app. This app contains a variety of puzzle games that challenge you to spell words based on clues. As you complete each game, you rack up points. The puzzles get increasingly difficult with each game.

What's more, you can create your own games and share them with friends and family members. You can create puzzle games based on different themes (see Figure 18-8). You can even use your camera to create game-screen backgrounds.

Figure 18-8:
You can create Spellista puzzles based on themes.

Chapter 19

Ten Things Not to Do with Glass

> *The world is full of magical things patiently waiting for our wits to grow sharper.*
>
> — Bertrand Russell

As soon as early adopters of Google Glass, called *Glass Explorers,* began to wear their Glass devices out and about, the computing media started to comment about how they saw Explorers using their devices — and especially about how *not* to wear Glass in various situations.

We've boiled down the variety of things not to do into ten helpful tips that should keep you from driving people crazy, not embarrassing yourself, or both.

Shower with It

This guideline was illustrated expertly by popular tech blogger Robert Scoble, an early tester and proponent of Glass who showed his enthusiasm by taking a picture of himself with his Glass on his head while he was taking a shower.

Though Scoble was discreet, the photo didn't exactly kindle enthusiasm among potential users. Instead, people (especially media people) took note that Glass doesn't tell people other than the wearer that the camera is on, so the device could encourage all sorts of mischief. Even Google's chief executive officer, Larry Page, remarked that he didn't appreciate the shower photo despite the publicity.

You can see Scoble's photo and read Page's response at `www.theverge.com/2013/5/15/4333656/larry-page-teases-robert-scoble-for-nude-google-glass-photo`.

What's more, exposing Glass to moisture could result in the device's becoming inoperative, because electronic circuitry and water mix as well as cola and pudding. The moral of this story is to keep your reputation intact and your Glass safe by ensuring that you keep the device away from your shower.

Take It on a Date

For a perfect example of how not to behave on a date while using Glass, see the video at `www.youtube.com/watch?v=8UjcqCx1Bvg`. If you'd rather just get a summary of the video and tips on what not to do with Glass on a date, here you go:

- Don't tell your date that you need Glass to see. Be honest about what it's for.
- Pay attention to your date instead of searching for jokes you can tell or getting the latest information about the TV series *Downton Abbey*.
- When you think the date is becoming boring, don't start watching sports highlights, play games, or take video calls.
- Your date may figure out what Google Glass is by asking Siri, Apple's personal assistant, on the iPhone attached to her head.

In summary, take Glass off. Instead, ask your date questions, and let him or her talk. You'll find out more about the other person and probably have a better time.

Pirate a Theatrical Movie

In addition to all the good things they do, smartphones allow people to record movies as they're playing in a theater and then post the movies on video-sharing websites such as YouTube.

Though smartphones haven't been banned in most movie theaters, the movie and theater industries has responded by offering $500 to theater employees who identify anyone recording movies with a smartphone, call the police, and contact the Motion Picture Association of America through an online form (www.fightfilmtheft.org).

Glass makes it even easier to record movies surreptitiously, because all you'd have to do is turn on the video-recording feature on Glass and then sit and watch the movie. You don't give away your intent by holding up your smartphone to watch the screen.

Though there haven't been any Glass bans at movie theaters at this writing, chances are you'll be turned away by a movie theater if an employee sees you sporting your Glass. Even if you put your Glass on after you enter the theater itself, theater workers will see the camera light that comes on while you're recording. Save yourself some aggravation, and don't be branded a movie pirate: Keep Glass in a safe place while you enjoy the movie.

Glass gets warmer the longer you use it. As you record a long video, the device may eventually get so hot that it overheats. The Glass circuitry likes excessive heat as much as any other electronic equipment does (that is, not), so unless you have extra money to spend on a new Glass device, overheating is another reason to keep Glass safe while you watch a movie.

Share Embarrassing Photos

A plethora of people still seem to think that taking naked pictures of themselves and sharing them with others is a good idea. The issue of "revenge porn," in which angry people post private photos publicly to embarrass their ex-boyfriends or ex-girlfriends, has prompted the introduction of laws against the practice, particularly in California.

Even when laws are in place, keep this axiom in mind: What goes on the Internet stays there forever. As with any other device you can use to share information, people are watching, so take care about what you share.

Sleep with It

Sometimes, it's hard to remember that you have Glass on your head because the device is so light and unobtrusive. If you don't remove it when you go to bed, however, and you like to move around during the night, you may find yourself with a broken Glass frame when you wake up. (The time when you wake up may vary, because you may hear the frame crunching.) You can also break the screen by putting pressure on the cube, which fits over your right eye.

If you haven't seen our explanation of every part of the Glass device, turn to Chapter 2. Then return here. Don't worry; we'll wait.

Even if you don't break the frame or screen, you may put so much pressure on the nose pads that your Glass won't fit properly on your face. The best-case scenario (if you can call it that) is that you may put pressure on the Camera button on top of the cube, take a bunch of pictures or videos, and then find the battery drained.

If you press the Camera button when your Glass is off, the device turns on and starts taking pictures and/or videos. The longer it takes pictures and videos, the shorter its battery life is.

Touch the Cube

The discussion of the cube in the preceding section leads to a corollary warning: Don't touch the cube. *Ever.* The cube is your screen, and it's fragile, so if you touch it, you run a high risk of breaking it. Google won't replace your Glass — even if it's under warranty — if you break the screen due to a brain freeze.

Break Rules — or the Law

You may have seen signs in businesses that say something along the lines of "We reserve the right to refuse service to anyone." The fact that you can wear your Glass almost everywhere doesn't mean that you *should* wear it everywhere, especially in a private business establishment. Even before Glass was available for sale to the public, one Glass Explorer was kicked out of a Seattle restaurant for refusing to stop using the device when the server asked him to stop (www.pcmag.com/article2/0,2817,2427804,00.asp).

Here's an important reminder for businesses: If Glass is banned in your business, post a reminder that people can see. That sign could save your business a lot of grief from angry patrons who were kicked out and complained online (or, worse, called lawyers).

Many casinos around the country have also banned Glass because of the high potential for collusion. One player at a blackjack table could see the dealer's cards through a Google Hangout on the Glass screen while his partner sits behind the dealer at a distance and uses his own Glass to see the dealer's cards.

Before you start using Glass, check with your state, province, and/or country about laws governing the use of driving while using a screen or other monitor. One shining example came in late October 2013, when a driver in California was cited for wearing her Google Glass while driving, thus violating a state law that requires people not to drive a vehicle equipped with a monitor unless it displays "global mapping displays, external media player (MP3), or satellite radio information."

If you have prescription shields affixed to your Glass, ask the business establishment for its policy on wearing Glass with prescription shields. You should strongly consider bringing a pair of regular prescription glasses with you, just in case the business doesn't allow patrons to wear Glass under any circumstances.

Try to Be a Fashionista

Glass isn't a means to show that you're "cool" by being on the cutting edge or by being superior to others. It's a device that's supposed to help enhance your life in certain situations. Using Glass when you're traveling around to get information is a good use of the device. If you're going to use Glass to correct other people or to act superior to them, don't be surprised to find people avoiding you. (If you're surprised by this statement, or if you feel the need to feel superior, we can't help you.)

Creep People Out

When you're talking with people, it's generally a good idea to take your Glass off your head to let them know that you're giving them your full attention. If you don't, the people you interact with may start behaving a bit oddly toward you. That behavior may be caused by your inability to maintain eye contact because you're too busy looking at the screen or moving your head up to activate your Glass. It may also be that people aren't sure that you're not taking their photo or taking a video surreptitiously. People *can* see that you're taking a photo or video, not only because the Glass screen lights up as you're doing so, but also because you have to speak a command or press the Camera button on top of the cube.

Show Off

Don't show off with Glass, either intentionally or unintentionally. You may have been annoyed by someone in a public place who talks on a smartphone. You'll be the person people love to hate if you make a phone call on your Glass, start speaking commands to it, or dictate an e-mail message it.

What's more, you shouldn't wear your Glass at a party or other gathering. It's nice to be able to check a person's name and interests before you talk to her, but she may be uncomfortable when you recite all her interests instead of asking her for information.

Glass does get attention, however, and people will ask you about it, so be gracious and answer questions. Find out how to share your Glass with other people safely by flipping to Chapter 7.

Chapter 20

Ten Ways to Make New Friends with Glass

You can't stay in your corner of the Forest waiting for others to come to you. You have to go to them sometimes.

— A. A. Milne, Winnie-the-Pooh

*W*hen you purchase Glass and start wearing it in public, you're going to get a lot of attention. Like it or not, Glass is an automatic conversation starter, and you're an ambassador for the device.

If you need to connect with other Glass users because you have a problem or issue and want to see how other users have handled it, or because you have a desire to connect with other Glass users, plenty of online and offline opportunities are available.

Connect with the Glass Community

The Glass Community is a private website you can access at `https://www.glass-community.com`. You need to sign in with your Google username and password.

The Glass Community has some basic rules that you should follow when you post messages:

- ✔ Make the text and any photos you include with your posts family-friendly.

- ✔ Be on your best behavior. The community has no tolerance for threats, harassment, or bullying.

- ✔ Avoid promoting yourself or your business. The website is for sharing information, not selling items or sending spam messages.

- ✔ Don't share your personal information with others.

- ✔ Include an interesting title with your post so it will attract readers.

- ✔ Provide complete background information in your posts, such as what you're doing with your Glass when an error occurs.

If you think that a post is inappropriate, the moderators make it easy for you to contact them and report any problems you think should be investigated.

Use Glass Hashtags on Social Networks

A *hashtag* is a word or a phrase that's preceded by a hash symbol (#) and identifies a post based on a specific topic. Hashtags are neat and convenient ways for people who want to find out more about a topic to search for it on social networking websites. If you search for #throughglass on Twitter, for example, you'll see posts by people who have sent text and/or photo tweets from Glass.

Google has established several hashtags you can add to your posts to make it easier for people to find your posts and start to follow you:

- ✔ **#throughglass:** When you send a post on a social networking website from your Glass, Glass adds this hashtag to the post automatically.

- ✔ **#ifihadglass:** This hashtag tells users about new ideas you have for Glass if you're a current user or just want to buy one.

- ✔ **#googleglass:** This hashtag notes that the post is related to Glass.

Move in Google+ Circles

The Google+ social networking website has a few communities where you can find current and past messages posted by other users so you can get feedback, find solutions, and meet other Glass users:

- **Glass Explorers:** If you're a Glass Explorer and want to chat about your experiences with other Explorers, visit the Glass Explorers community at `https://plus.google.com/u/0/communities/107405100380970813362`.

- **Glass Owners:** If you own a Glass, the Glass Owners community is the place to get the latest information, ask questions, and meet other owners. Visit this community at `https://plus.google.com/u/0/communities/105306970516704274456`.

- **Project Glass by Google:** Here, you can interact with other Glass hardware and app developers. Just visit `https://plus.google.com/u/0/communities/103179668665582410083`.

 You can also search for Glass communities on Google+ by clicking or tapping the Home button and then clicking or tapping Communities. In the Communities screen that appears, type the community name you're searching for (such as *Glass*) in the Search for Communities box.

Hook Up with LinkedIn Groups

If you frequent the LinkedIn professional networking website, you can join a couple of Glass groups so you can interact with other Glass users around the world. Here are two that you may be interested in:

- **Google Glass Developers:** This group is for Glass app developers. You can exchange ideas, post information about your latest app project, and get the latest news about Glass and Glass development. Access this group at `www.linkedin.com/groups/Google-Glass-Developers-4872136/about`.

- **Google Glass Forum:** The Google Glass Forum group at `www.linkedin.com/groups/Google-Glass-Forum-5135187/about` discusses app ideas and innovative uses of Google Glass, and hosts general discussion of how Glass will affect the lives of its users and society in general.

LinkedIn limits the number of groups you can join to 50. If you already belong to 50 groups, you won't be able to add Glass groups unless you leave an existing group.

You can search for Glass groups in LinkedIn. You'll see a long list of groups that you can join on the results page.

Wear Glass in Public

Glass is a built-in conversation starter. People will ask you about the device and what it can do. They may be curious, or they may be afraid that you're going to do something nefarious with your Glass.

Be gracious. Answer all questions as completely as you can. You may want to find the answers on your Glass while you're talking to that person — but be sure to tell that person you're going to look for answers on Glass before you do it.

Do Show-and-Tell with Glass

When you get together with family members, friends, or colleagues who are interested in technology, be sure to bring Glass with you and answer questions. You can even share your Glass if you feel comfortable doing that. Maybe people will be inspired to buy a Glass themselves . . . and perhaps even buy this book!

Some people may be put off by, if not openly derisive of, Glass and wearable computers in general. See Chapter 19 for things *not* to do with Glass and how to present the device properly.

Develop Something Great for Glass

If you enjoy tinkering with new technologies and want to see what they can do, Glass is a great platform for you. You get to work on something exciting and new, and you may also get to develop the next big thing (and earn some extra income).

To find out more about "hacking" your Glass by using the Glass Development Kit and other development tools, see Chapter 14.

Get Your Glass at a Base Camp

Google has set up Base Camps in several cities around the United States, including San Francisco, New York, and Los Angeles. If you reserved your Glass on the Glass website (www.google.com/glass), you can pick up the device at one of these Base Camps and also do the following cool stuff:

- ✔ Get your Glass fitted with the help of a Glass Guide.
- ✔ Find out more about Glassware, Glass hardware, and cool things you can do with the device.
- ✔ Chat with Glass Guides and other Glass users.

If you purchase Glass at a Base Camp, the sales tax for your state applies to the purchase. If you purchase your Glass in San Francisco, for example, your purchase price includes California sales tax at the San Francisco tax rate.

Attend Glass Events

Google+ communities announce in-person events about Glass in various communities around the world, so visit these communities often to get the most recent information.

The Meetup website (www.meetup.com) is a great place to visit for the latest events, not only about Glass, but also about wearable devices in general. On Meetup, you can search for groups and/or events, and you can change the search radius from your current location, such as within 100 miles of San Francisco.

Host Your Own Glass Event

If you can't find an event, start one of your own! You can create events and/or post messages asking whether people would be interested in the following places:

- ✔ Forums on social networking websites such as LinkedIn and Google+
- ✔ Event sites such as Meetup
- ✔ Chamber-of-commerce and other business sites

Index

• *H* •

• *I* •

About the Authors

Eric Butow: Eric Butow began writing books in 2000, when he wrote *Master Visually Windows 2000 Server* (IDG Books). Since then, he has written or co-written 23 other books. He lives in Jackson, California, and is the owner of Butow Communications Group, an online marketing improvement firm.

Robert Stepisnik: Robert Stepisnik specializes in developing software and lives in Radlje ob Dravi, Slovenia. He has always had a passion for being a pioneer in new areas of computing technology, and Google Glass is no different. He is the owner of Stepisnik Software, which offers custom software development for wearable technologies.

Dedication

Eric Butow: For my mother, who sets a good example every day.

Robert Stepisnik: For my amazing girlfriend, Natalija, who is always there for me and is making me a better person. For my father, who left this world too soon. And for my mother, who is always cooking something good. .

Authors' Acknowledgments

Eric Butow: My thanks to my family and friends for their boundless support. I want to thank my awesome literary agent, Carole Jelen, as well as Kathy Simpson and Katie Mohr for everything. Finally, I want to thank my co-author, Robert Stepisnik, because without him, this book would not be in your hands.

Robert Stepisnik: A big thanks to my family, friends, and everyone who made this book possible, especially my co-author, Eric Butow. Also, I want to thank Google for giving me a chance to be part of history as one of the first Glass Explorers.

Wiley Publishing would like to thank Mark LaFay for the use of his Google Glass for photos of the device and for his expert input as technical reviewer.

Publisher's Acknowledgments

Acquisitions Editor: Katie Mohr

Project Editor: Kathy Simpson

Copy Editor: Kathy Simpson

Technical Editors: Mark LaFay, Robert Scoble

Editorial Assistant: Anne Sullivan

Sr. Editorial Assistant: Cherie Case

Project Coordinator: Erin Zeltner

Cover Image: © Wiley / Michael E. Trent

Apple & Mac

iPad For Dummies,
5th Edition
978-1-118-72306-7

iPhone For Dummies,
7th Edition
978-1-118-69083-3

Macs All-in-One
For Dummies, 4th Edition
978-1-118-82210-4

OS X Mavericks
For Dummies
978-1-118-69188-5

Blogging & Social Media

Facebook For Dummies,
5th Edition
978-1-118-63312-0

Social Media Engagement
For Dummies
978-1-118-53019-1

WordPress For Dummies,
6th Edition
978-1-118-79161-5

Business

Stock Investing
For Dummies, 4th Edition
978-1-118-37678-2

Investing For Dummies,
6th Edition
978-0-470-90545-6

Personal Finance
For Dummies, 7th Edition
978-1-118-11785-9

QuickBooks 2014
For Dummies
978-1-118-72005-9

Small Business Marketing
Kit For Dummies,
3rd Edition
978-1-118-31183-7

Careers

Job Interviews
For Dummies, 4th Edition
978-1-118-11290-8

Job Searching with Social
Media For Dummies,
2nd Edition
978-1-118-67856-5

Personal Branding
For Dummies
978-1-118-11792-7

Resumes For Dummies,
6th Edition
978-0-470-87361-8

Starting an Etsy Business
For Dummies, 2nd Edition
978-1-118-59024-9

Diet & Nutrition

Belly Fat Diet For Dummies
978-1-118-34585-6

Mediterranean Diet
For Dummies
978-1-118-71525-3

Nutrition For Dummies,
5th Edition
978-0-470-93231-5

Digital Photography

Digital SLR Photography
All-in-One For Dummies,
2nd Edition
978-1-118-59082-9

Digital SLR Video &
Filmmaking For Dummies
978-1-118-36598-4

Photoshop Elements 12
For Dummies
978-1-118-72714-0

Gardening

Herb Gardening
For Dummies, 2nd Edition
978-0-470-61778-6

Gardening with Free-Range
Chickens For Dummies
978-1-118-54754-0

Health

Boosting Your Immunity
For Dummies
978-1-118-40200-9

Diabetes For Dummies,
4th Edition
978-1-118-29447-5

Living Paleo For Dummies
978-1-118-29405-5

Big Data

Big Data For Dummies
978-1-118-50422-2

Data Visualization
For Dummies
978-1-118-50289-1

Hadoop For Dummies
978-1-118-60755-8

Language &
Foreign Language

500 Spanish Verbs
For Dummies
978-1-118-02382-2

English Grammar
For Dummies, 2nd Edition
978-0-470-54664-2

French All-in-One
For Dummies
978-1-118-22815-9

German Essentials
For Dummies
978-1-118-18422-6

Italian For Dummies,
2nd Edition
978-1-118-00465-4

e **Available in print and e-book formats.**

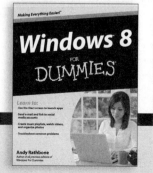

Math & Science

Algebra I For Dummies,
2nd Edition
978-0-470-55964-2

Anatomy and Physiology
For Dummies, 2nd Edition
978-0-470-92326-9

Astronomy For Dummies,
3rd Edition
978-1-118-37697-3

Biology For Dummies,
2nd Edition
978-0-470-59875-7

Chemistry For Dummies,
2nd Edition
978-1-118-00730-3

1001 Algebra II Practice
Problems For Dummies
978-1-118-44662-1

Microsoft Office

Excel 2013 For Dummies
978-1-118-51012-4

Office 2013 All-in-One
For Dummies
978-1-118-51636-2

PowerPoint 2013
For Dummies
978-1-118-50253-2

Word 2013 For Dummies
978-1-118-49123-2

Music

Blues Harmonica
For Dummies
978-1-118-25269-7

Guitar For Dummies,
3rd Edition
978-1-118-11554-1

iPod & iTunes
For Dummies, 10th Edition
978-1-118-50864-0

Programming

Beginning Programming
with C For Dummies
978-1-118-73763-7

Excel VBA Programming
For Dummies, 3rd Edition
978-1-118-49037-2

Java For Dummies,
6th Edition
978-1-118-40780-6

Religion & Inspiration

The Bible For Dummies
978-0-7645-5296-0

Buddhism For Dummies,
2nd Edition
978-1-118-02379-2

Catholicism For Dummies,
2nd Edition
978-1-118-07778-8

Self-Help & Relationships

Beating Sugar Addiction
For Dummies
978-1-118-54645-1

Meditation For Dummies,
3rd Edition
978-1-118-29144-3

Seniors

Laptops For Seniors
For Dummies, 3rd Edition
978-1-118-71105-7

Computers For Seniors
For Dummies, 3rd Edition
978-1-118-11553-4

iPad For Seniors
For Dummies, 6th Edition
978-1-118-72826-0

Social Security
For Dummies
978-1-118-20573-0

Smartphones & Tablets

Android Phones
For Dummies, 2nd Edition
978-1-118-72030-1

Nexus Tablets
For Dummies
978-1-118-77243-0

Samsung Galaxy S 4
For Dummies
978-1-118-64222-1

Samsung Galaxy Tabs
For Dummies
978-1-118-77294-2

Test Prep

ACT For Dummies,
5th Edition
978-1-118-01259-8

ASVAB For Dummies,
3rd Edition
978-0-470-63760-9

GRE For Dummies,
7th Edition
978-0-470-88921-3

Officer Candidate Tests
For Dummies
978-0-470-59876-4

Physician's Assistant Exam
For Dummies
978-1-118-11556-5

Series 7 Exam For Dummies
978-0-470-09932-2

Windows 8

Windows 8.1 All-in-One
For Dummies
978-1-118-82087-2

Windows 8.1 For Dummies
978-1-118-82121-3

Windows 8.1 For Dummies,
Book + DVD Bundle
978-1-118-82107-7

Available in print and e-book formats.

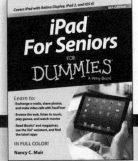
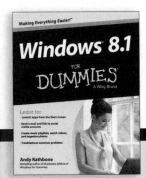

Available wherever books are sold. **For more information or to order direct visit www.dummies.com**

Take Dummies with you everywhere you go!

Whether you are excited about e-books, want more from the web, must have your mobile apps, or are swept up in social media, Dummies makes everything easier.

For Dummies is the global leader in the reference category and one of the most trusted and highly regarded brands in the world. No longer just focused on books, customers now have access to the For Dummies content they need in the format they want. Let us help you develop a solution that will fit your brand and help you connect with your customers.

Advertising & Sponsorships

Connect with an engaged audience on a powerful multimedia site, and position your message alongside expert how-to content.

Targeted ads   s sponsorship

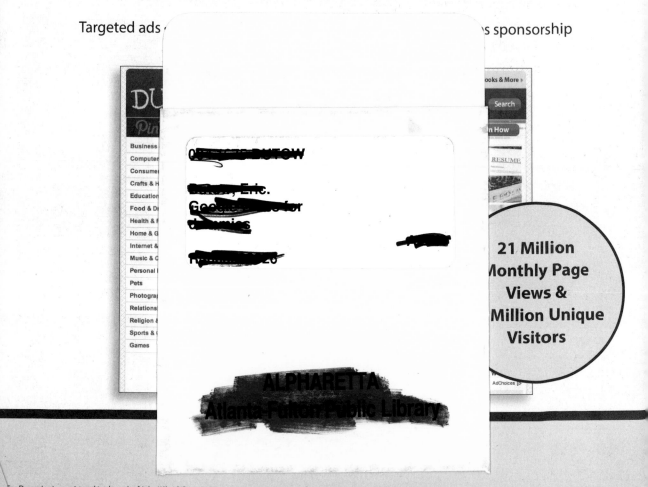

21 Million Monthly Page Views & Million Unique Visitors